A SIMPLIFIED GUIDE TO USING STATISTICAL TECHNIQUES WITH COMPUTER APPLICATIONS

Margaret N. Morrison
Director of Computer Services
Armstrong State College
Savannah, Georgia

Prentice-Hall, Inc. Englewood Cliffs, New Jersey

Prentice-Hall International, Inc., *London*
Prentice-Hall of Australia, Pty. Ltd., *Sydney*
Prentice-Hall Canada, Inc., *Toronto*
Prentice-Hall of India Private Ltd., *New Delhi*
Prentice-Hall of Japan, Inc., *Tokyo*
Prentice-Hall of Southeast Asia Pte. Ltd., *Singapore*
Whitehall Books, Ltd., Wellington, *New Zealand*
Editora Prentice-Hall do Brasil Ltda., *Rio de Janeiro*
Prentice-Hall Hispanoamericana, S.A., *Mexico*

To Good Friends
Slowly made, to last a lifetime

Library of Congress Cataloging-in-Publication Data

Morrison, Margaret N.
A simplified guide to using statistical techniques—with computer applications.

Includes index.
1. Mathematical statistics—Data processing.
I. Title.
QA276.4.M68 1986 519.5 86-3242

ISBN 0-13-810185-X

Printed in the United States of America

Contents

Preface

Most professionals would like to make greater use of statistics, but they are not always sure where, when, and how to apply them. *A Simplified Guide to Using Statistical Techniques* was conceived as a result of extensive contact with people for whom statistics was a mystery. The same sense of mystery was felt even by people who had been exposed to statistics, who had worked the formulas, and perhaps even used the results. This book seeks to help anyone overcome statistical ignorance and fear. The reader need only follow the explanations and examples drawn from various practical applications in different fields—business, industry, education, health, and politics.

The book is useful for all those who wish to apply statistical techniques to obtain answers to their questions—for managers and administrators who want to understand fully what their statistical research people are telling them; for students who wish to improve their understanding by supplementing their standard statistics text books; for people who find themselves required to use or to teach statistics with outdated training; for people who have aspirations to be managers; and for any professionals who desire to use the computer in developing statistics.

A Simplified Guide to Using Statistical Techniques is exactly that—simplified. You don't have to learn math symbols to understand computer statistics; computer printers normally do not have math symbols. There are no statistical tables in the back of this book; therefore, you don't have to remember whether to go across the table or down the table. The computer produces the figure you would have searched a table to find. There are no coin tossing or card drawing examples such as are found in many statistics books.

Statistical jargon will not be used for the statistics presented. Instead, the book will explain and enlighten through the extensive use of practical examples, drawn from real-life situations. Throughout this book you will find actual applications and interpretations of statistics to solve very real day-to-day problems.

A Simplified Guide to Using Statistical Techniques is divided into two broad parts. Part 1 explains the current nature and use of statistics:

statistics that describe the nature of the items being counted and statistics that use the items for prediction.

The second part of the book outlines several studies as they would be produced by a computer. Each chapter in Part 2 guides you in the use of what was learned in Part 1. Most important, you will be shown how the printed results may be interpreted. Although we will concentrate on interpreting the data the computer provides in these applications, much valuable general statistical information can be obtained with the simplified approach. You are not expected to become a computer programmer or a statistician, but it should be easier for you to obtain the answers to your questions.

Acknowledgments

My appreciation goes to the following people for giving suggestions and support, or allowing me to use examples in which they directed the studies: Robert Magnus, Grace B. Martin, Gerald Sandy, Madeline Sandy, Thomas Kemp, William Megathlin, Andrew Mazzoli, David Ferris, Benjamin Thomas, Mickie Harbin, Morris Simpson, Clifford Burgess, Jeff Dukes, John L. Stegall, Robert A. Burnett; to George E. Parker, Executive Editor, Business and Professional Books Division of Prentice-Hall, Inc., for his encouragement and professional management; to my staff: Janice Christy, Rivanna Dudley, Greg Simpson, Tommy Hann, Susan Lucas, Jay Gardner, who took up the slack every time I took a day of vacation to work on this book (they may claim there was not that much "slack"); and to my family.

A Word from the Author on This Book's Practical Value

Statistics pertain to anything that can be counted—defective items manufactured; cells or bacteria; deaths from certain diseases; results of certain treatments; answers given on surveys or tests; wages; sales; profits; costs. The use of statistics either gives a valid representation of the items counted or misrepresents and obscures the nature of the items counted. A computer is "dumb" in producing counts of anything and everything; it falls to the human to control what should be counted.

You may be interested in developing statistics but even more interested in understanding the results. Many people require statistics to make sound judgments but really don't understand enough to take full advantage of the statistical answers to a problem. For example, if a store manager received a statistical report showing that the daily receipts averaged $23,000 over the past year, that manager might draw the conclusion that an additional investment in the business, which has been considered, is in order. However, if the manager had understood some of the other figures presented, he or she might have seen that a figure representing the variation was inordinately high—$20,000. This figure indicates that the receipts, although averaging $23,000, ranged from $3,000 to $43,000; the manager might then reconsider the conclusion and begin to look at the causes of such a great variation. Understanding all the figures presented can keep you from drawing the wrong conclusions, help you to reevaluate the basic question, and lead you to the correct, most profitable, conclusion.

Many errors with statistics develop because of the failure to interpret the numbers produced in the proper manner. The problems of statistical use are noticeable even among people who have had statistical training. Knowledge of formulas does not necessarily mean proper application to the problem at hand. People often develop solutions to the wrong problem—and then alter the problem to fit their results. This manipulation of numbers has led to a general mistrust of statistical answers.

In this book you will find techniques to formulate questions for which statistical answers *can* be developed. When a question has been properly formulated (for example, is there a significant difference between my sales this year and those of last year?), application of an appropriate statistical procedure is facilitated. The approach to question formulation is aimed at the statistical procedures discussed in the book, but it is broad enough to be applied to the use of any statistical procedure.

Even if a result has been formulated properly, the presentation of the result can lead to misinterpretation. Great care must be taken to present all assumptions, decisions, and changes that were used to obtain the result. Even a useful statistical result such as the government's cost-of-living index has been questioned, in part because of the failure to present it correctly; factors used to obtain the index are not known, change frequently, or are not of comparable value.

You, with the help provided by this book, will be confident about presenting what you are trying to present. The helter-skelter approach taken by some people often leads to incorrect or incomplete answers. Some people blindly apply any statistic to any items in the hope of finding some answers that can be accepted. The advent of statistical computer packages has aggravated this problem. A computer statistics package is a set of computer programs, or software, designed to produce statistics with a minimum amount of effort and computer knowledge on the part of the user. These packages are so easy to use, and can so quickly produce page after page of statistics that you might be tempted to go hunting for possible connections that don't exist. In this book you will be given techniques to develop your problem logically and soundly.

Computers make it easy to use statistics. In today's world, not using statistics, like not using computers, means being left behind. Let's begin to look at how we can use both to get ahead.

How to Use This Book

This book is meant to be a *learning* book, not a "look it up and do it this way" book. Knowing the *why* will make the *how* come much more easily; the aim is to *lead* the reader to learn. When learning occurs, people can go beyond this book in using the power of a computer to help make decisions or draw conclusions based on statistics. With learning in mind, the book is written in a logical-flow format, progressing from simple statistics to more difficult statistics, using a step-by-step, building-from-the-bottom-up method. Therefore, it is recommended that you read the book in its entirety before getting started with statistics, before using it as a reference book.

When you are reading this book, illustrations unrelated to your field of interest should *not* be skipped. Examples have been chosen that can be understood by everyone; nonbusiness people can understand simple records about number of sales and profit; nonacademics can understand grade-point average and school subjects. The end result is the same—understanding the characteristics and general function of the statistic illustrated.

For your use as a reference, a glossary has been included and, in Part 1, the first reference to a specific statistic is in bold print. The concepts presented in Part 1 are repeated in the real-life examples in Part 2; you learn by repetition. These examples are combined with related areas in each chapter, so that you can use them later as reference material.

Even if you don't have access to a computer, the book is useful as a beginning book on statistics. If you use a computer you will surely have some familiarity with your computer equipment, and at least a little knowledge of statistics, before buying or using a computer statistics package. Knowledge of specific statistics software and specific equipment can be obtained from the manuals provided when you purchase that software or equipment. A little knowledge of statistics and computers in general can be obtained from this book. Although you do not expect to learn statistics by some miracle, or magic, if you have a need or at least a mild interest, this book gives you the opportunity to *read and learn* statistics presented in simple language.

SECTION

I

UNDERSTANDING BASIC STATISTICS

1

Starting with Statistics

1. USING FACTS

Have you ever noticed how so many intelligent people can make the simplest things difficult? You want to say: get to the point—the bottom line. The bottom line for businesses is (1) how much do we owe?, and (2) how much is owed us? Numbers are money. Where numbers are not directly related to money, but to services or improvements or research, the same two questions can be asked. This time the emphasis is on the first question: how much can we do to be more efficient, to provide better service, to improve quality?

In everyday living, we are exposed to numbers in the form of statistics probably more than in any other form. Now, national data bases on computers make it possible to access all kinds of information by phone or other communications devices. The amount of information in the form of data and statistics is endless, yet many people know very little about the uses of statistics.

Even if your organization is large enough to have a statistician, it would be more efficient for you to gain some knowledge of statistics; then you are in a better position to get the information you need, instead of what someone else thinks you need. A few managers may be reluctant to talk to the statistician because the managers know nothing about statistics, so they correspond through a middle man. This can be ineffective (did you ever play "Gossip" as a child?). If you want to base your decisions on facts, you must be willing to spend some time getting *all* the facts possible, or the result might be a solution to the wrong problem—or a solution to a problem that doesn't exist. At times, some facts can be worse than no facts.

In your experience, you may have seen people who go off half-cocked and make a decision before getting all the facts, or base their decisions on opinions presented as facts. The use of numbers, combined with your best judgment, will help in making these decisions. The numbers may do nothing more than confirm that what you think, is, in fact, true. The numbers may lessen the uncertainty if some is present, as it is in many decisions. You can weigh the *cost* of making a wrong decision against the *risk* of making a wrong decision. Sometimes getting the facts will be much easier than you might imagine; other times, getting the facts together will be like putting Humpty Dumpty together again—not worth the time and cost. However, with computers the time and cost is minimized.

Whether you are getting the facts yourself or having someone else gather them, all aspects of the problem must be well defined. Correcting what is thought to be one problem might cause five other problems, or valuable time might be wasted collecting unnecessary facts. Whenever possible, you should decide what your actions will be before getting the results—*if* result, *then* solution. Some oversimplified examples are: (1) for inventory, if stock is less than fifty items, then reorder; (2) for quality control, if articles are flawless, then sell as best quality with money-back guarantee; (3) if there are some flaws, then sell as seconds; or (4) if there are many flaws, then discard. We will have to define "some" and "many." Knowing beforehand what the alternative solutions are will help in defining the problem.

The approach taken in this book is to learn by practical example, not by definition. You will find examples and explanations primarily of the following questions:

1. For what kinds of things do you collect data?
2. How much data do you collect?
3. What statistics do you use?
4. How do you interpret the results?
5. What decisions do you make about the results?
6. What are the chances of making the right decisions?

The favorite toy of some children is an old box of some kind. The point is: we don't have to be fancy; keep it simple until such time as we are ready for the more sophisticated. In the beginning, you may be a little off base, but at least you won't be way out in left field. The goal of this book is to motivate you to get started with statistics, either for the first time ever, or for the first time lately.

2. CLASSIFYING DATA

If you've ever had a statistics course and fallen asleep during the lecture on this topic, you have missed the key to practical statistics. The appropriate statistics are completely dependent on the type of data used. Before you can begin to use statistical methods, you have to know what kind of data you have. Statisticians usually classify data into four different groups, called levels. There is very little difference between some of these, and we will use only two groups—labeled data and continuous data. Those of you who are more serious about statistics can find the shades of difference and the fancier names in most statistics books.

Labeled data is that which can be categorized and counted. **Continuous** data can be measured, and arithmetic can be used on it. Examples of the types of data in Table 1–1 will make the differences between the types of data obvious. Statistics to use with each type, along with examples, will follow later in the book.

All of the data items in Table 1–1 can be called **variables;** the value will vary between different people, or cases, or observations. Sex is a variable that can have only two different valid values. However, if the group from which the data is gathered contains only males, then sex would not be considered a variable; it would be considered a **constant.**

Suppose a toothpaste manufacturer does a survey on how many people prefer metal toothpaste tubes to plastic toothpaste tubes. The data items are labeled; they are metal or plastic. The statistics obtained are counts of how many people prefer metal and how many people prefer plastic. The result of the survey gives two numbers, each in a different category: a count for the labeled item, metal, and a count for the labeled item, plastic. If we consider the numbers of tubes of toothpaste sold, then we are using continuous, or measurement, data because numbers sold can have any value, which can constantly change as more tubes are sold. We can add to the numbers sold and find a total number or an average number sold per day, per week, etc. In the education field, the result of a count of students taking or not taking a particular course is considered labeled data—yes or no. If we are considering the grades, in number form, made by the students taking the course, we have continuous data.

Continuous Data	Labeled Data
Age	Sex
Height	Race
Weight	Birthplace
Time	Genetic Characteristics
Distance	Marital Status
Speed	Occupation
Strength	Geographic Areas
Temperature	Political Preference
Capacity	Religious Preference
Physiological Rates	Blood Type
Calories	Vitamin Type
Education Level	Brand
Ranks	Make
Scores	Model
Grades	Ratings
Population	Low/Medium/High
Income	Defective/Not Defective
Sales	For/Neutral/Against
Losses	True/False
Prices	Yes/No/Undecided
Absences	Success/Failure
Dividends	Urban/Rural
Taxes	

Table 1-1 Types of Data

Specific examples will be used with the two different types of data. Even though the data items might be different for your work or interest, keep in mind that the only difference is in the items. *The types of data, continuous and labeled, and the appropriate statistics will be the same in every field.* Also remember that anything that can be counted or measured can be considered statistics and can be used to produce other statistics.

The purpose of this book is to provide some knowledge of statistics for those people who know the other details of their work or interest; to provide an overview of the possibilities of computer-produced statistics for the nonstatistician—exposure to collecting and using data, coding for the computer, reading computer-produced tables, and interpreting simple statistics.

2

How Does the Data Look

1. COUNTING MADE EASY BY COMPUTER

When we are interested in finding out *how many* we use a **frequency table** (or **frequency distribution**). This refers to arranging data in order in a list to show how many times each single value occurred. The frequency table's main purpose is to give a first general impression of how your data looks. Each value can be used, as in test scores, with every occurrence of a score being listed in the value column and the number of people making that score in the frequency column.

When we have a large number of values, it may be difficult to work with the data or to see easily where the values are concentrated. In such a case, large sets of data may be grouped, and assigned the number for each group, such as the number of people making A, the number making B, etc. A usual rule is to arrange data into groups with the groups being the same width and not overlapping. That is, each group will span the same number of values, such as 0–9, 10–19, 20–29,. . . , and a value can be included in only one group. However, with some data like test scores (less than 60 = F), one or both ends may be all-inclusive. This may be necessary when there are large gaps in the data. The rule about not overlapping should be followed. A score should be included in only one group. The order of listing depends upon the type of data. When it is done manually it is usually arranged from best to worst; for test scores the arrangement would be from high to low score (descending order), and with golf scores the arrangement would be from low to high (ascending order), as the low score is the best. Computer statistics packages usually use ascending order, regardless. Some packages may provide a choice of different arrangements for different tables.

With a frequency table you can easily see the lowest value, the highest value, the value or values with the largest count (called the **mode**), or the value or values with the smallest count, and the difference between the lowest and highest (called the **range**).

When we want to know the percentage of occurrences for each value, we look at the **relative frequency** in which the percentage in each group is given. In many cases the percentage is the figure we want to quote. This is needed when we are comparing two or more separate sets of data, particularly those that have a different total number of cases; for example, comparing the effects of different types of treatment recorded for 1,000 lower animals with the same recorded for 100 humans, or the results of tests for 800 men compared to those for 500 women. Of course, what you are comparing has to be related data—the same treatments or the same tests. Table 2-1 shows the frequency table with a relative frequency column and a cumulative frequency column. The **cumulative frequency** column is simply the addition, one by one, of each individual figure in the relative frequency column. We can imagine that these are job levels. To generalize, they are numbered, with 1 as the lowest level and 7 as the highest level. A glance at the table shows 41.7 percent of employees in the lowest level job. In the upper levels, 5, 6, and 7, 14 percent is determined by adding the relative frequency for those designations or by subtracting 86.0 (in the cumulative frequency column) from 100.0. You could also have requested the computer to combine values 5, 6, and 7 before producing the table if this is what you wanted to see.

Value	Frequency	Relative Frequency Percent	Cumulative Frequency Percent
1	1135	41.7	41.7
2	406	14.9	56.6
3	413	15.2	71.8
4	386	14.2	86.0
5	202	7.4	93.4
6	151	5.5	98.9
7	30	1.1	100.0
Total	2723		

Table 2-1 Frequencies with Percentages

Tables 2-2 and 2-3 show the same data, but separated by plant for comparison. We will assume that Table 2-2 is the main plant and Table 2-3 is the branch plant. The percentage differences in job levels 1 and 2 at the two plants may need investigation. The main plant may have too large a percentage promoted to level 2, or the branch plant may have too small a percentage promoted, or the percentage may be a verification that job levels are in the proportion they were meant to be in.

In the cumulative frequency column in Table 2-3, the total is printed as 100 percent. However, if we added: 49.3 + 10.9 + 15.2 + . . ., we would not get 100 percent. The minor discrepancy here, and similar ones throughout the book, can be explained by a simple example. 1.06 + 1.06 = 2.12 which is 2.1 rounded to one decimal; 1.1 + 1.1 = 2.2 if we round before adding. The 49.3, obtained by dividing 497 by the total of 1009, was actually 49.257 rounded to three decimal places.

This type of information can be obtained on almost any data that is a part of your computer files, such as age, seniority, sex, race, sick days taken, departments, supervisors, salaries, and what's selling and what's not selling (turnover in merchandise). Maybe there are patterns that are not obvious until the data is assimilated in some order.

For frequency tables with the data having a label rather than a numeric value (section 1.2), of course the data cannot be grouped: surveys with answers like always, usually, sometimes, never; count

Value	Frequency	Relative Frequency Percent	Cumulative Frequency Percent
1	638	37.2	37.2
2	296	17.3	54.5
3	260	15.2	69.7
4	247	14.4	84.1
5	163	9.5	93.6
6	95	5.5	99.1
7	15	.9	100.0
Total	1714		

Table 2-2 Frequency Table: Job Level—Main Plant

Value	Frequency	Relative Frequency Percent	Cumulative Frequency Percent
1	497	49.3	49.3
2	110	10.9	60.2
3	153	15.2	75.3
4	139	13.8	89.1
5	56	5.6	94.7
6	39	3.9	98.5
7	15	1.5	100.0*
Total	1009		

*Total is not exactly 100 percent.
This is because of computer rounding.

Table 2-3 Frequency Table: Job Level—Branch Plant

of employees by job title rather than job level; number of students by state of residency; or different types of treatments.

If the necessary information is there, you can receive data grouped in any way that makes sense. The computer does not know if it makes sense; that is left to you. Although it is still simplified, let's look at a practical example in which the data, as is, cannot be grouped; a supplier of parts does not stock certain parts because of low demand in the past. When customers request those parts, the company special-orders them. An example of this type of list is shown in Table 2-4. Special-ordering means paying extra freight costs and a delay to the customer. It has been determined that if a certain part has to be special-ordered at least five times in one year, then the company will begin

Part Number	Number of Times Special Order
A	2
B	8
C	3
D	1
E	5
F	4

Table 2-4 Special Orders

stocking that part. Since the special orders should have been entered into the computer files, a frequency table of these will show, in Table 2–4, that parts B and E will need to become part of the inventory. Alternatively, a list of frequencies of 5 or greater could be obtained if you did not want the whole list.

2. SHOWING THE DATA IN A GRAPH

When precise figures are not necessary, and to get a clearer picture of how the data compares, a frequency distribution can be shown as

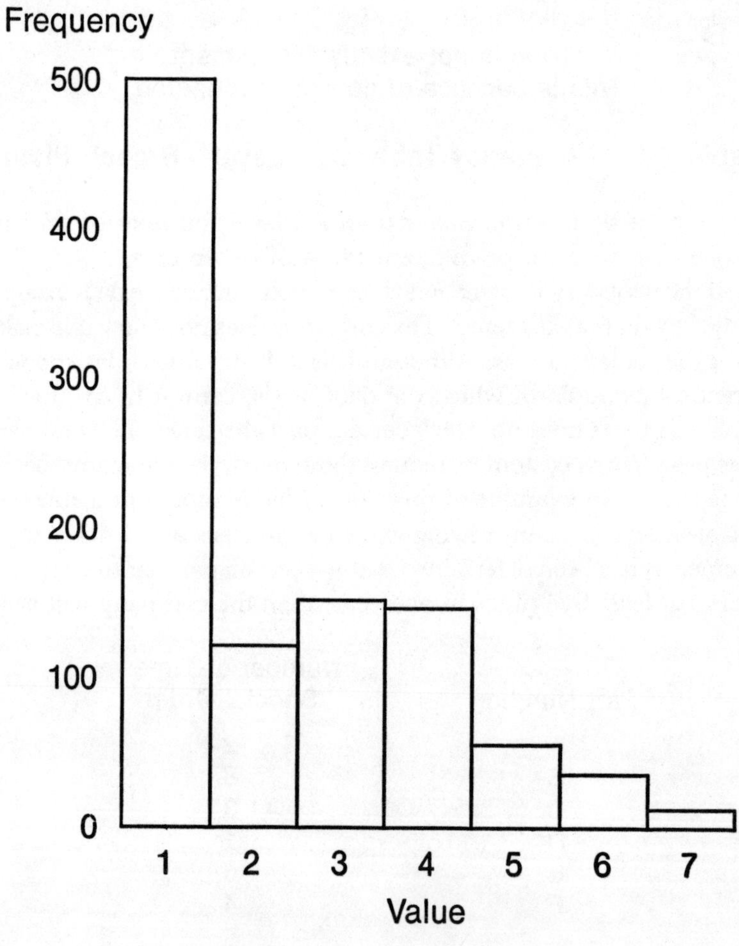

Figure 2-1 Job Level

a **histogram** or **bar graph.** The lengths of the bars indicate the number for each value or the number in each group of values. Figure 2-1 shows the data used in Table 2-3 in histogram form. Some statistics packages show this type of graph with horizontal separated bars as in Figure 2-2, or with vertical separated bars. Usually, grouped data is more appropriate for these graphs, especially if there is much data and the range of data is wide.

Table 2-5 shows a frequency table ungrouped and then the same data grouped in Table 2-6. In the ungrouped Table, 2-5, the **minimum** is 200 and the **maximum** is 950. The RANGE is 750 (950 − 200). The **mode** is 400, as the largest frequency is 20, which belongs to that value. After you see the ungrouped data, or at least know the minimum and maximum, you may want to change the way the groupings are made. In this example for Table 2-6, all groups are integers and are the same width (199). When you want to provide for decimal data, the group would include, for example, 200–400. When you reach 400.0, the data falls into the next group. With a few simple commands to the computer you may receive the data grouped some other way. Much of this depends on whether you're

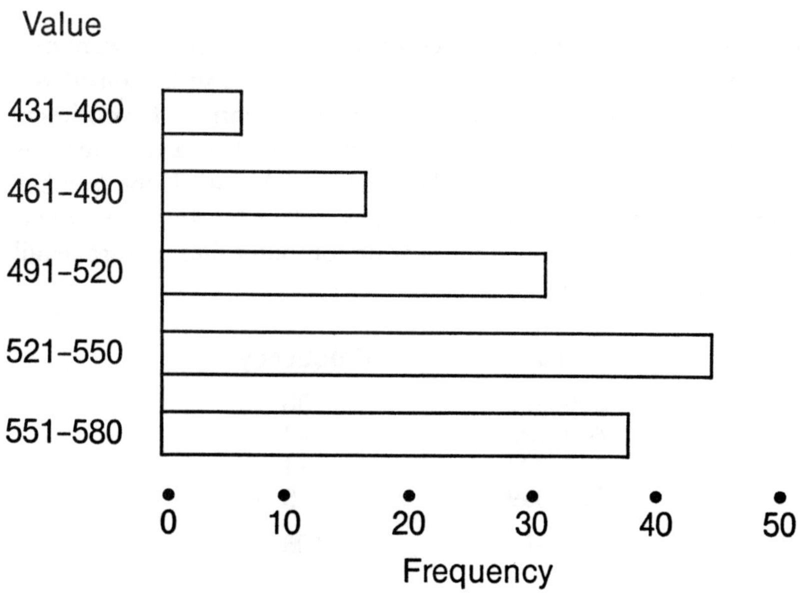

Figure 2-2 Horizontal Bar Chart with Grouped Data

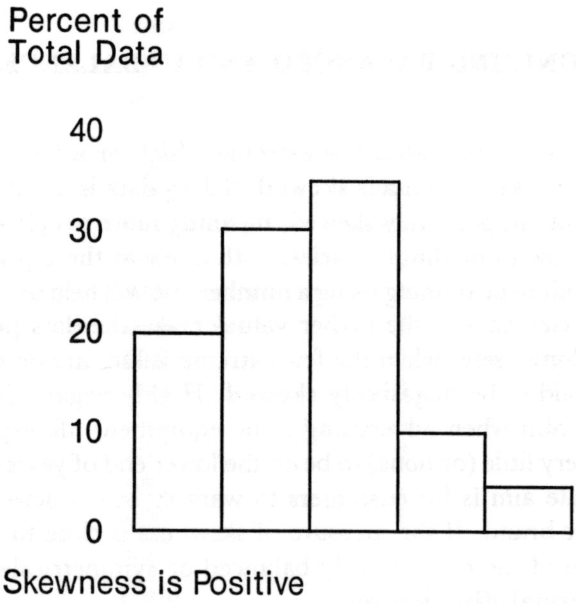

Figure 2-4

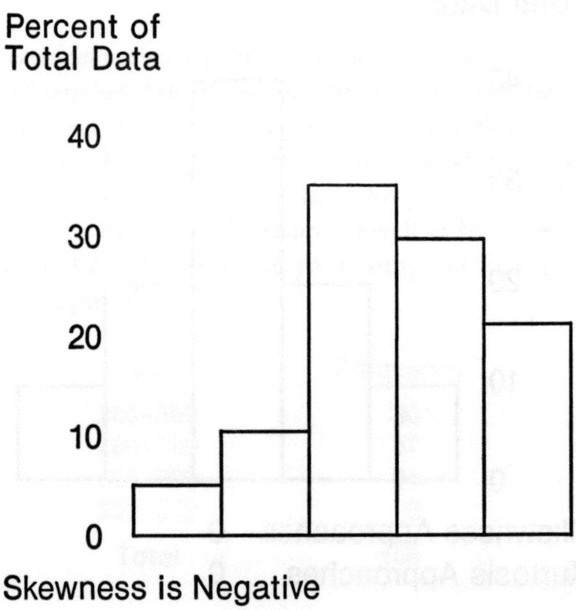

Figure 2-5

Another lesser used measure of distribution is **kurtosis.** This refers to the relative peakedness or flatness of the data when shown in a histogram or other type of graph. If the measure of kurtosis is close to zero, as with skewness, the distribution of the data approaches a normal distribution. A positive measure of kurtosis means peaked and narrow; negative means flatter and wider; and the farther from zero, the more pronounced these characteristics are. Both measures are useful only with continuous data (section 1.2). Figures 2–3 through 2–7 show graphs or pictures of how the data would look, generally, with these different characteristics. They are not meant to be exact pictures but only to give you a clearer idea of some measures of distribution. Figure 2–6 shows fairly consistent data; that is, 60 percent of the data is clustered together and another 30 percent (15 per-

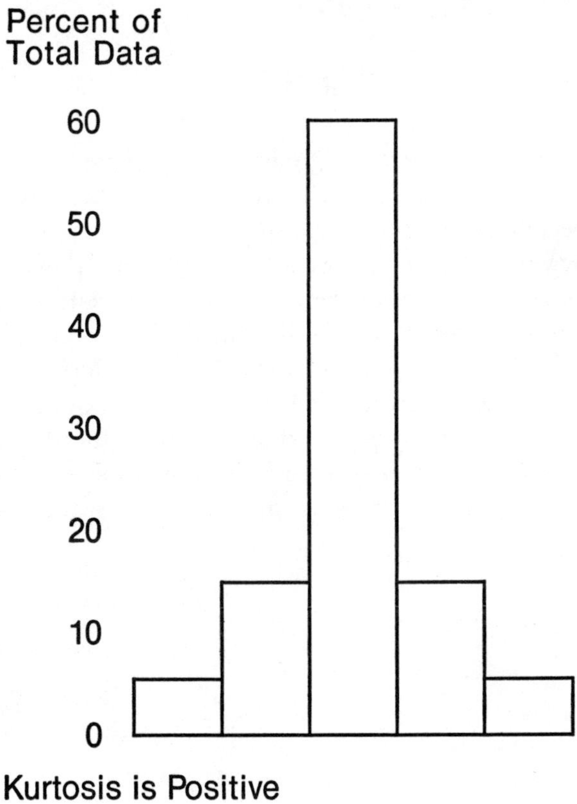

Figure 2-6

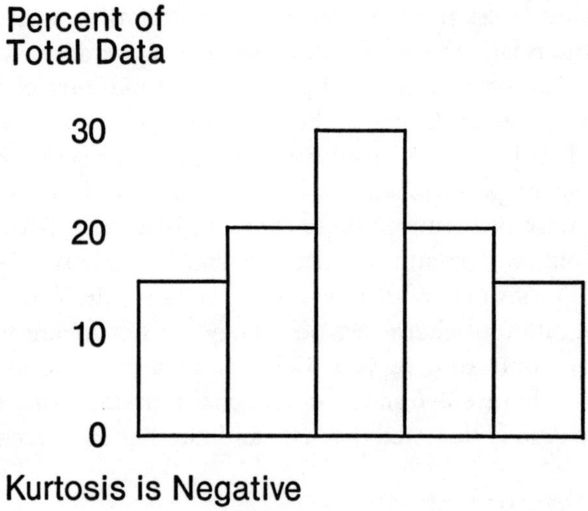

Figure 2-7

cent on each side) is close to this group. This is desirable when precision is necessary, that is, when measurements should be close to a standard or expectation. Figure 2–1 also shows data that is positively skewed (only a few values on the upper end), and Figure 2–2 shows data that is negatively skewed (fewer values on the lower end).

A computer statistics package can easily produce counts and graphs of anything and everything in your computer files in any way you want it counted or combined. If that's not enough information for you, then you may enter additional data for this purpose—obtaining frequency tables, percentages, and graphs—statistics mentioned here and in the following chapters.

3

How to Describe the Data

1. WHAT IS AVERAGE?

While the frequency table shows you what your data looks like individually, one statistic used to describe the data as a whole is the **average.** Many people, including professionals, are impatient about looking at arrays of figures. Just give them a number they can quote when the need arises; this one number quite often is the average. The most commonly used averages are the mean average, the median average, and the mode. A student was overheard criticizing some teachers: "They are so hard that the *average* student can't even make an A," not realizing how the comment sounded—that by most definitions of average, the grade of an *average* student would be C, not A.

The most widely used average is probably the arithmetic **mean,** and it is what is usually referred to when we say average. This is obtained by summing the values or scores and dividing by the total number of values or scores. Everyone is familiar with the mean—averaging a set of test scores to predict a final grade in school classes, or knowing average temperature, average miles per gallon, average salary, stock market average, average life of equipment, or average yards gained in football. The result may not necessarily be one of the numbers in the set; the average may not exist as in some descriptions of the average family with 2.2 children. For example, given ten numbers,

$$1, \ 1, \ 1, \ 2, \ 2, \ 2, \ 3, \ 4, \ 4, \ 5,$$

whose sum is 25, the mean is 25/10 or 2.5.

The halfway point (or middle point or center) of a set of data, when arranged in order, is called the **median.** If there is an even number of frequencies, the median falls between the two center values.

For example, given ten numbers,

1, 1, 1, 2, 2, 3, 3, 4, 4, 5,

the median falls between the fifth and sixth value (between 2 and 3) which, in this case, makes the median 2.5. Consider the following ten numbers:

1, 2, 2, 2, 2, 2, 3, 3, 3, 3.

We would pick a 2 as the median because the fifth value is 2 and the sixth value is 2. Some computer statistics would count the number of 2s nearer 1 and the number nearer 3, and give the median as a decimal; that is, if the "/" divides the set in half:

1, 2, 2, 2, 2, / 2, 3, 3, 3, 3,

we have more 2s nearer 1 than 3, thus the computer may declare the median to be 1.99 or 1.98 instead of 2.

The value that appears most frequently is the **mode** (section 2.1)—the most popular brand or the most popular size, the customer's preference or need. Clearly, this is the average to use when ordering merchandise to sell. A set of data may have more than one mode. A computer determination of the mode is usually limited to one of these, probably the first one encountered in a list arranged in ascending order. It is necessary to look at the frequency table (section 2.1) to find any others. For example, given ten numbers,

1, 2, 2, 2, 3, 3, 3, 4, 5, 5,

the computer would list 2 as the mode even though there is an equal number of 3s. There may be no mode if all frequencies are equal.

Value	Frequency	Relative Frequency Percent	Cumulative Frequency Percent
20	1	2.9	2.9
21	7	20.0	22.9
22	19	54.3	77.1
23	7	20.0	97.1
24	1	2.9	100.0
Total	35		
	Mean 22	Median 22	Mode 22

Table 3-1 Frequency Table—Normal Distribution

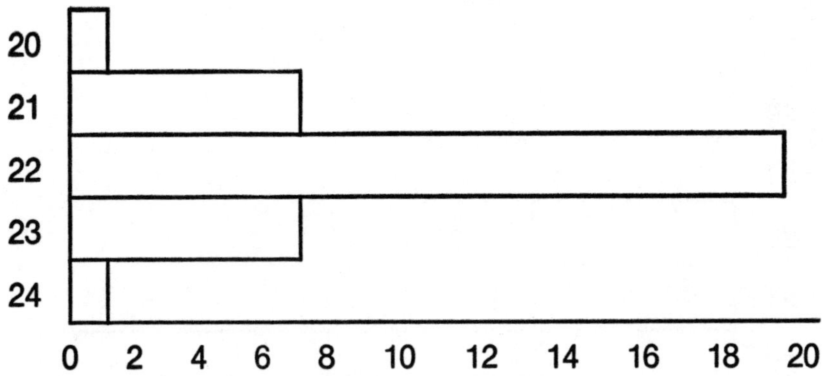

Figure 3-1 Graph of Normal Distribution

Table 3-1 and Figure 3-1 show a limited set of data illustrating a normal distribution (section 2.3). The mean is 22, obtained by adding each value multiplied by its frequency, and dividing by the total of 35. The median (middle) value is the eighteenth value, which is contained within the frequency count for 22. The mode, most frequent, is 22. The graph, Figure 3-1, is symmetrical. Therefore, skewness (section 2.3) is exactly 0, and all averages are the same. The peak in the center produces a small positive kurtosis (section 2.3). When a graph of data is close to symmetrical, the mean, median, and mode will be close, and skewness will be close to 0.

Table 3-2 and Figure 3-2 show negatively skewed data because

Value	Frequency	Relative Frequency Percent	Cumulative Frequency Percent
13	1	2.1	2.1
14	2	4.3	6.4
20	6	12.8	19.1
21	7	14.9	34.0
22	18	38.3	72.3
23	7	14.9	87.2
24	6	12.8	100.0
Total	47		

Mean 21.5 Median 21.9 Mode 22

Table 3-2 Frequency Table—Skewed (Negative)

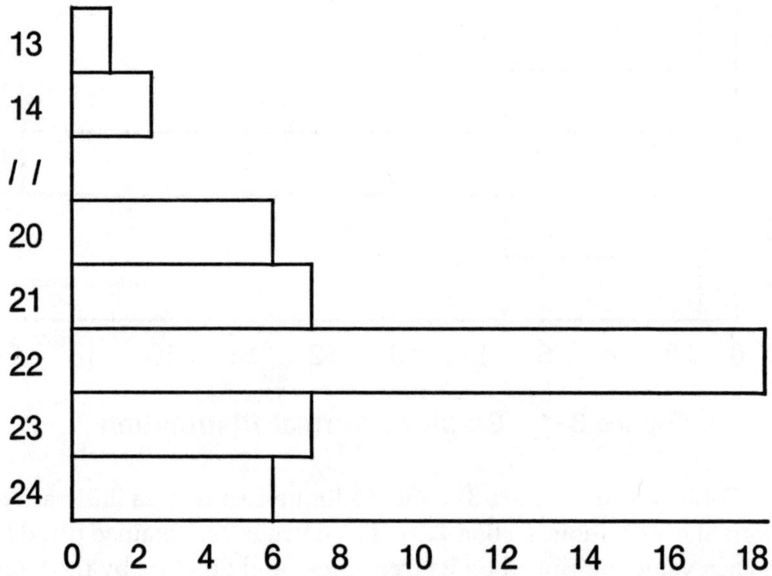

Figure 3-2 Graph of Skewed Distribution (Negative)

13 and 14 contains three relatively extreme data points. The "//" in Figure 3–2 denotes a gap in the data on the integer scale. Kurtosis is still positive with the peak at 22. In Table 3–2, notice that the mean is less than the median and mode for this set of negatively skewed data. For the median we would pick 22, which contains the twenty-fourth (middle) value from the total of 47. The computer figured it as 21.9 when considering the relationship of all the data.

Table 3–3 and Figure 3–3 show positively skewed data because 29 and 30 contain two relatively extreme data points on the upper end. Kurtosis is again positive. Notice that the mean is now larger than the median and mode for the set of positively skewed data. As the data set is small for illustrative purposes, the difference is very small.

Table 3–4 and Figure 3–4 show data that results in a negative kurtosis—the graph is relatively flat. If the counts in the frequency column are fairly close, as in Table 3–4, the graph will appear fairly flat, as in Figure 3–4. Skewness is very close to 0; there are no extreme values. The mean and median are close because of no extreme values.

Value	Frequency	Relative Frequency Percent	Cumulative Frequency Percent
20	6	13.0	13.0
21	7	15.2	28.3
22	18	39.1	67.4
23	7	15.2	82.6
24	6	13.0	95.7
29	1	2.2	97.8
30	1	2.2	100.0
Total	46		

Mean 22.3 Median 22.1 Mode 22

Table 3-3 Frequency Table—Skewed (Positive)

For an idea of low and high skewness, the relative values are shown in modified frequency distributions in Figure 3-5a through Figure 3-5d. Each dot represents one unit for the value shown. In Figure 3-5a, skewness is 0 because the distribution is symmetrical; there are no extreme values. With the addition of the extreme value

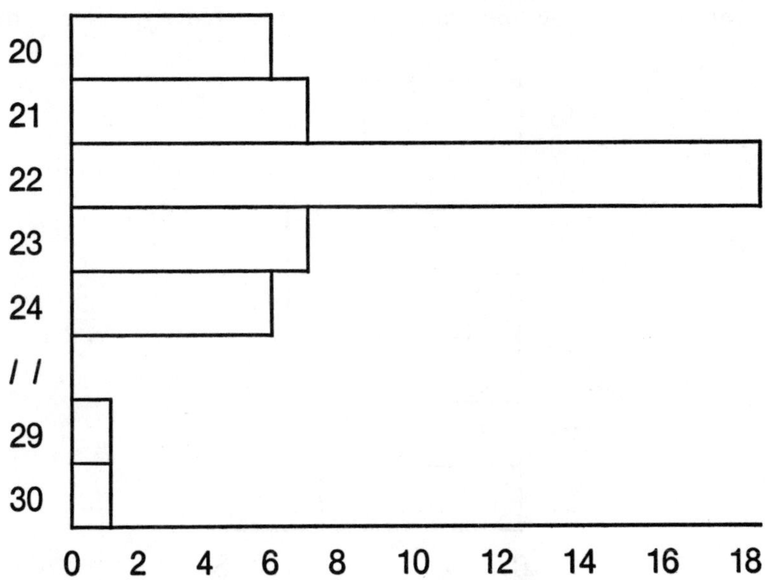

Figure 3-3 Graph of Skewed Distribution (Positive)

Value	Frequency	Relative Frequency Percent	Cumulative Frequency Percent
18	4	11.8	11.8
19	5	14.7	26.5
20	6	17.6	44.1
21	6	17.6	61.8
22	7	20.6	82.4
23	1	2.9	85.3
24	5	14.7	100.0
Total	34		

Mean 20.9 Median 20.8 Mode 22

Table 3-4 Frequency Table—Kurtosis (Negative)

of 100,000 in Figure 3-5b, notice how the skewness increases. Next, more data is added on the lower end for Figure 3-5c. The skewness increases again because 100,000 now is considered even more atypical in relation to the amount of data clustered together on the lower end. Finally, more data is added on the lower end for Figure 3-5d. As more cases become clustered together, the result is that one extreme case (or a few extreme cases) is definitely considered atypical and the

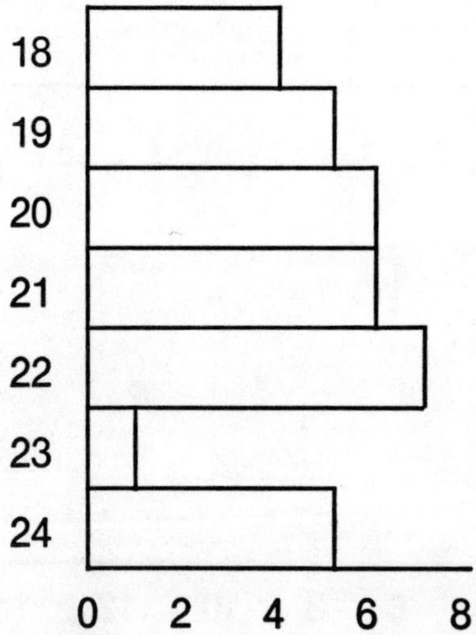

Figure 3-4 Graph Showing Kurtosis (Negative)

Figure 3-5a Skewness = 0

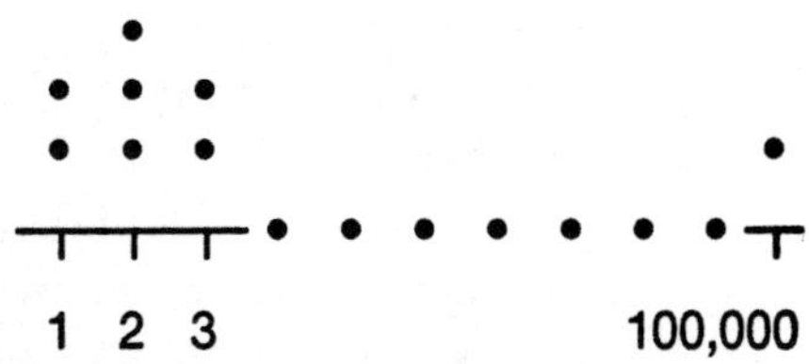

Figure 3-5b Skewness = 2.8

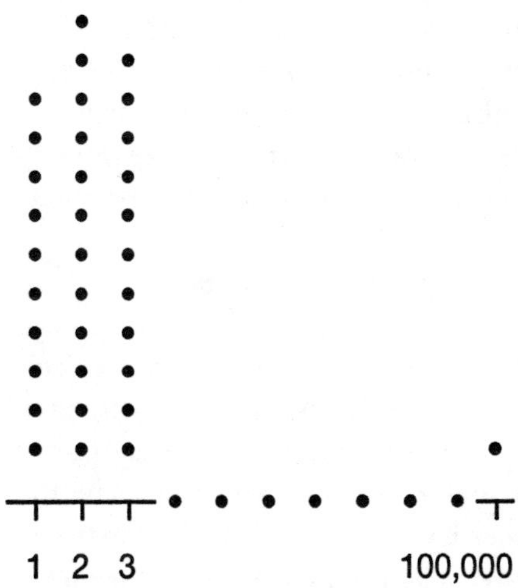

Figure 3-5c Skewness = 5.8

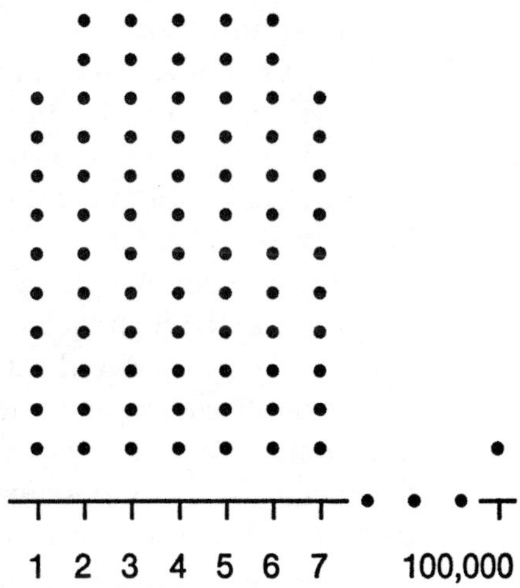

Figure 3-5d Skewness = 9

measure of skewness increases. If the extreme value were on the left in the graph, the effect of skewness would be the same, but it would be negative.

2. WHAT AVERAGE TO USE?

If the mean average sales of salespersons are known, then the total sales can be figured by multiplying those averages by the total number of salespersons. Continually aiming for higher total sales, we can reverse the above and figure what each salesperson's average should be. This is not true of the median average or the mode. They have no further statistical value beyond giving a quick description of the characteristics of a set of data, and in some cases this is all we need. As we shall see later, the mean average has many uses statistically. It can be used to produce other statistics necessary for many decisions or conclusions.

The mean is affected by extremely high or extremely low values (skewness), therefore it may be misleading to base decisions solely on the mean average. The result may be an average with no meaning. The basketball coach, knowing the average height of the next opponent, also needs to know if they have an extremely tall center. As the mean is affected by extremely high values or extremely low values, it is not a good statistic to depend on solely when the data is skewed considerably. In a distribution that is positively skewed, the mean is the highest of the three averages, and if it is negatively skewed, the lowest of the three averages, as you saw in the previous tables of this chapter.

The median (middle value) is not affected by extreme values such as the few in the high income group at the top. It is the best average to use to describe skewed data—that is, if you do not want your decision to be affected by extreme values. This is also true when you have a set of data for which you know the extreme values are missing, but for which you need those extreme values to accurately describe the data. Psychological studies sometimes have this problem. The extreme cases may not be available for study. The median is sometimes used with test scores when wanting to determine percentiles, in particular the 50th percentile. If a measurement is in the 50th percentile, 50 percent of the scores are below that measurement; i.e., a percentile indicates where one measurement stands in relation to the group of measurements. It is the percentage of measurements that fall below a measurement; for example, if 70 percent of the measurements are below a measurement of 85, then the measurement of 85 is in the 70th percentile.

The mode (most frequent) also is not affected by extreme values. It is used when describing the typical case rather than the more commonly used average case. The typical family may have 2 children whereas the average family may have 2.2 children. The mode, if all frequencies are not equal, is the only one of the three which will always be a member of the set of data (if all values occur with the same frequency, we have no mode). When you are considering data that is labeled, the mean and median have little use but the mode can be used: More people replied yes to a survey question than replied no or don't know; more packages of brand X were sold; treatment A gave positive results in more cases than any other treatment; fertilizer F produced more bushels per test plot. The mode is the most representative average for this type of problem.

The computer does not know that some averages have no use, that you cannot average the number of sales of different brands together and have a number that means anything;

brand X - 3000 units sold
brand Y - 4000 units sold
brand Z - 5000 units sold

The mean average determined by the computer is brand Y. Again, the mode of 5000 is probably the statistic of interest rather than the fact that brand Y is average. We are assuming that brands X, Y, and Z are similar items at comparable prices. In the next section we shall look at a way to compare items that are not similar. It is left to the person to use those computer results that are appropriate to the problem.

3. HOW FAR FROM AVERAGE?

We mentioned the range—the difference between the lowest and highest values. This is a simple measure of variability in data but it considers only the highest and lowest values. For our basketball team, we may know the height of the tallest player but still know nothing about the heights of the other players. A measure of variability may be thought of as a measure of consistency. Does a machine always produce the exact size part needed or does it vary? For this example, the range may be all that is needed. If the smallest or largest parts will not fit the equipment for which they were manufactured, the range is too large. In most cases another measure of variability is more useful—one that considers all the data and how much variability is in it. The **standard deviation** is commonly used. This takes into consideration how far from (mean) average each value is.

Let's use the game of golf to make it easier to comprehend the concept of standard deviation. If you think of par as average (the mean), the actual golf scores can be represented as par, one above par, one below par, etc. On a number line this could be represented, with par 70, as follows:

```
--I---I---I---I---I---I---I--
 67  68  69 PAR71  72  73
 -3  -2  -1  0   1   2   3
```

In this example, the standard deviation is shown as 1 (67 to 68, 68 to 69, etc.). The lower numbers are called **z-scores.** The score of 73 would be 3 above par or a z-score of +3. The score of 68 would be 2 below par or a z-score of −2. Now, the farther from par (or mean average), the larger the number, disregarding the negative. We will assume a reasonable limit on the size of the number. A group of accomplished golfers would all be close to par; there would be little variability; they would be consistent.

This way of looking at data, as above or below average and by how much, is the basis of using standard deviation. As in the golf scores, the mean average is determined and the data then is described in standard deviations. This is the actual distance from the mean. However, in order to compare unlike sets of data, a method is used in which the mean is always 0 and the data is divided into six parts. Anything beyond these six parts is extreme.

Let's look at an example using data that ranges from 40 to 100 with a mean of 70 and separated into six parts. The standard deviation is 10.

```
--I---I---I---I---I---I---I--
 40  50  60  70  80  90  100
 -3  -2  -1   0   1   2   3
```

Let's look at another example using data ranging from 70 to 100 with a mean of 85, separated into six parts. Now the standard deviation is 5.

```
--I---I---I---I---I---I---I--
 70  75  80  85  90  95  100
 -3  -2  -1   0   1   2   3
```

The first example shows more variability. It is obvious that a set of data with a standard deviation of 10 is more variable, or not as consistent as a set of data with a standard deviation of 5. Of course, very little data is as simple as these examples have been made. The changing of the standard deviation was only for illustration. A set of data has one calculated standard deviation that does not change unless data is added to or deleted from the set.

If you have not already recognized this, in the first example, a value of 98 would have a z-score of 2.8 and this is 28 points above the average of 70. This figure is also considered 2.8 standard deviations (2.8 times the standard deviation of 10 gives the value 28). Knowing the mean and standard deviation (S.D.), you can compute the z-score for a given value, X.

$$z = \frac{X - \text{mean}}{\text{S.D.}}$$

$$z = \frac{98 - 70}{10}$$

$$z = \frac{28}{10}$$

$$z = 2.8$$

More difficult values we will leave to the computer to figure for us. For the z-score, we need only remember that the closer to 0, the closer to the average, and the farther from 0, the farther from average, either above or below.

The standard deviation is used somewhat like percentage to make comparisons. Both are a way of comparing by splitting into groups. Companies may hire employees who are in the top 10 percent of their class. This is, in effect, obtained by splitting the class into ten equal parts, or groups, after arranging the class in descending grade order (highest to lowest). However, standard deviation is based on the theory that much of the data will be close to the mean average with less and less data as values get farther from the mean. A large amount of data is assumed to approach this normal distribution.

We took the liberty of explaining the concept of standard deviation by splitting into six groups because, generally, over 99 percent of data will fall within these six groups, or +3 and −3 standard deviations. Approximately 68 percent will fall within +1 and −1 standard deviations and approximately 95 percent will fall within +2 and −2 standard deviations. If the data is skewed, these percentages may not be exact. For example: Given a mean of 70 and a standard deviation of 10, 68 percent of data should be between 60 and 80 which is

70 − 1(10) and 70 + 1(10);

95 percent of data should be between 50 and 90 which is

70 − 2(10) and 70 + 2(10);

99 percent of data should be between 40 and 100 which is

70 − 3(10) and 70 + 3(10),

as shown in the next modified number line.

```
--I---I---I---I---I---I---I--
 40  50  60  70  80  90 100
 -3  -2  -1   0   1   2   3
          I-68%-I
      I-----95%-----I
  I---------99%---------I
```

The standard deviation is a way of comparing data that is measured in the same units, and having approximately the same mean. We can say a standard deviation of 10 shows more variability than a standard deviation of 5, i.e., a set of data with a larger standard deviation is more variable, or not as consistent as another set of comparable data with a smaller standard deviation.

Before continuing this discussion, let's stop and study some examples. The data in Table 3–5 could be representative of many different problems. We will suppose it represents units sold during a certain sales period by 70 salespersons. It is read: two salespersons sold 240 units, two sold 250 units, four sold 260 units, and so forth. Interpretation of Table 3–5:

1. Skewness of .9 shows a slight positive skewness—a few salespersons, but very few, did extremely well, and probably none did extremely poorly. The mean is greater than the median and the mode for positive skewness.
2. Kurtosis of 1.18 denotes a slight peak—some concentration of sales around 300 and 310.
3. Mean of 322 and standard deviation of 50 can be evaluated thus: Those salespersons selling between 272 and 372 units are within one standard deviation of average.
   a. 322 − 1(50) = 272
   b. 322 + 1(50) = 372

Units Sold	Number of Salespersons	Relative Frequency Percent	Cumulative Frequency Percent
240	2	2.9	2.9
250	2	2.9	5.7
260	4	5.7	11.4
270	3	4.3	15.4
280	5	7.1	22.9
290	3	4.3	27.1
300	9	12.9	40.0
310	10	14.3	54.3
320	4	5.7	60.0
330	6	8.6	68.6
340	1	1.4	70.0
350	5	7.1	77.1
360	3	4.3	81.4
370	5	7.1	88.6
380	2	2.9	91.4
400	1	1.4	92.9
410	1	1.4	94.3
420	2	2.9	97.1
450	1	1.4	98.6
490	1	1.4	100.0
Total	70		

Minimum	240	Median 312
Maximum	490	Mode 310
Range	250	Mean 322 (rounded)
Skewness	0.9	S.D. 50 (rounded)
Kurtosis	1.18	(A span of 50 is 1 S.D.)
Std. Error	6	C.V.% 15
95% C.I.	310-334	

Table 3-5 Frequency Table—Sales Analysis

Those selling between 222 and 422 units are within two standard deviations of average.

c. 322 − 2(50) = 222

d. 322 + 2(50) = 422

Those selling between 172 and 472 units are within three standard deviations of average.

e. 322 − 3(50) = 172
f. 322 + 3(50) = 472

These calculations, 3.a. through 3.f., are shown as a modified bar graph in Figure 3-6. Each bar represents the width of one standard deviation, and the number of salespersons (from Table 3-5) falling within each group is shown for the frequency portion.

4. Above average:
   within +1 S.D. from 322 to 372 units (3.b.)
   between +1 and +2 S.D. from 372 to 422 units (3.d.)
   between +2 and +3 S.D. from 422 to 472 units (3.f.)

   Looking again at Table 3-5 or Figure 3-6, we see one person above 472 at 490, or more than 3 S.D. above average. This one did extremely well. (We already guessed that we had at least one because of the slight positive skewness of .9) Those eleven in the group from 380 to 450 were much above average, or more than + 1 S.D.

5. Below average:
   within −1 S.D. from 272 to 322 units (3.a.)

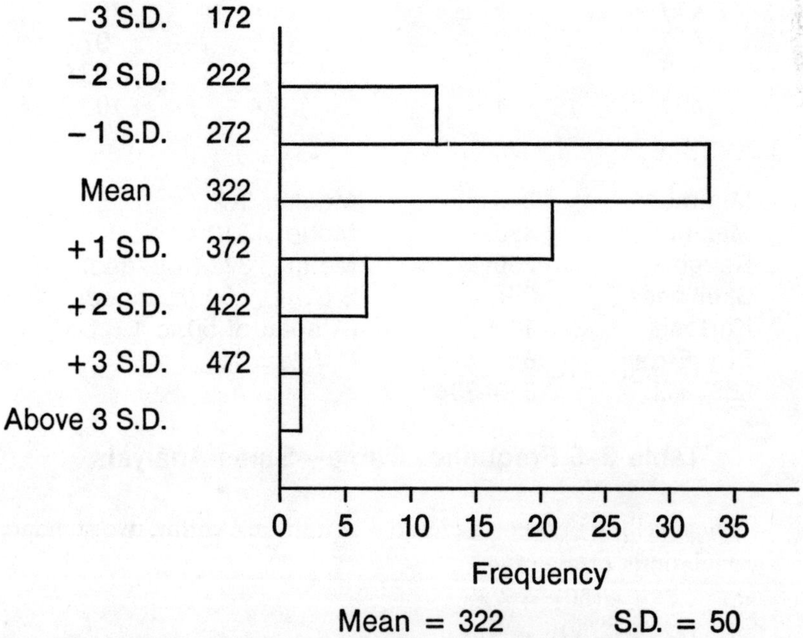

Figure 3-6 Modified Bar Graph Illustrating Standard Deviation for Sales Analysis

between −1 and −2 S.D. from 222 to 272 units (3.c.)
between −2 and −3 S.D. from 172 to 222 units (3.e.)
In this example, no salesperson was considered poor, although the eleven listed for the numbers 240 to 270 were below average. The cumulative frequency column in Table 3–5 shows those eleven to be 15.4 percent of the total. None were less than −2 S.D. (none less than 222).

The Coefficient of Variation (C.V.%), Standard Error (Std. Error), and 95 percent Confidence Interval (95% C. I.) will be explained later in this chapter.

Table 3–6 is compiled for the same salespersons for a different sales period. Looking at the frequency table only, with the statistics included, we can guess that the sales record was not as good. The minimum is less; the maximum is less with the same range. Checking the other figures, we know it is not as good. All averages (median, mode, and mean) are less than those in Table 3–5. The standard deviation is larger, denoting more variability or less consistency, although it is negligible (53 compared with 50 in Table 3–5). Further interpretation of Table 3–6:

1. Skewness of .47 shows less positive skewness; fewer or no salespersons did extremely well.
2. Kurtosis of −.04 means no peaks in data—no concentration of data around adjacent values.
3. Mean of 303 and standard deviation of 53 can be calculated and compared as we did for Table 3–5.
4. Above average:
   within +1 S.D. 303 to 356 (303 + 1(53) = 356)
   between +1 and +2 S.D. 356 to 409 (303 + 2(53) = 409)
   between +2 and +3 S.D. 409 to 462 (303 + 3(53) = 462)
   As we guessed, no salesperson was extremely good, none above 3 S.D. which was 462. Only two were above 2 S.D., which was 409. The group above 356 were much above average.
5. Below average:
   within −1 S.D. 250 to 303 (303 − 1(53) = 250)
   between −1 and −2 S.D. 197 to 250 (303 − 2(53) = 197)
   between −2 and −3 S.D. 144 to 197 (303 − 3(53) = 144)
   Again, no salesperson was considered poor and none were below −2 S.D.

Units Sold	Number of Salespersons	Relative Frequency Percent	Cumulative Frequency Percent
200	2	2.9	2.9
220	1	1.4	4.3
230	2	2.9	7.1
240	3	4.3	11.4
250	4	5.7	17.1
260	6	8.6	25.7
270	5	7.1	32.9
280	8	11.4	44.3
290	3	4.3	48.6
300	2	2.9	51.4
310	7	10.0	61.4
320	8	11.4	72.9
330	2	2.9	75.7
340	3	4.3	80.0
350	1	1.4	81.4
360	3	4.3	85.7
370	2	2.9	88.6
380	2	2.9	91.4
390	2	2.9	94.3
400	2	2.9	97.1
420	1	1.4	98.6
450	1	1.4	100.0
Total	70		

Minimum	200	Median	300
Maximum	450	Mode	280
Range	250	Mean	303 (rounded)
Skewness	0.47	S.D.	53 (rounded)
Kurtosis	– 0.04	(A span of 53 is 1 S.D.)	
Std. Error	6	C.V.%	17
95% C.I.	290–316		

Table 3-6 Frequency Table—Sales Analysis 2

Using the calculations in items 4 and 5 for the data in Table 3-6, the standard deviation and number of salespersons is shown in Figure 3-7. It is set up like Figure 3-6. For Table 3-6, the cumulative frequency column (section 2.1) can be used to determine who is in the

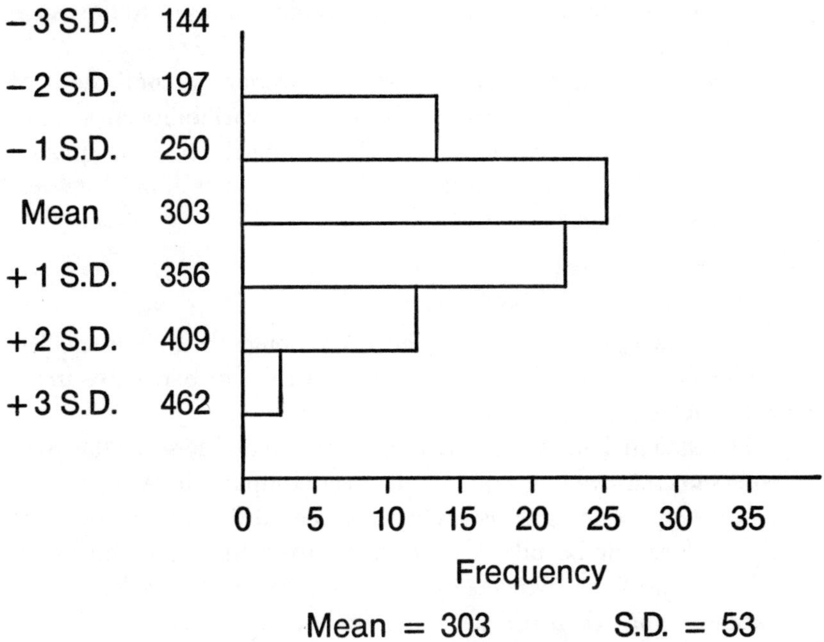

Figure 3-7 Modified Bar Graph Illustrating Standard Deviation for Sales Analysis 2

top 10 percent of sales, or in the bottom 10 percent. The bottom 10 percent is approximately 240 or less (cumulative frequency of 11.4 at the value 240) and the top 10 percent is approximately 380 or above (cumulative frequency of 91.4 at the value 380). The 51.4 percent at the value 300 indicates that over half were below average, below the mean of 303.

Suppose, however, that we are dealing with sets of data that are not comparable—average productivity where production time per unit or per job is different, height in inches compared to weight in pounds, score on two unlike tests. If a person received a score of 75 on one test and the same person received a score of 80 on another test, we would assume the second score was better. If the 75 was a z-score of .5 and the 80 was a z-score of −1, then we know the score of 75 was, in fact, relatively better than the score of 80. Comparing actual scores or separate sets of raw data may tell us nothing when we need to make comparisons of unlike data. This is the use of the z-score, since it is not relative to the size of the original values or to the range of the data.

Computer statistics packages should produce a **coefficient of variation (C.V.)**. This shows the amount of variability in a set of data as a percentage. Actually, it is the standard deviation divided by the mean. Since results were rounded in the sales analysis tables, the figures may not be exact.

Example 1: S.D. = 5, C.V. = 40%
Example 2: S.D. = 11, C.V. = 29%

There is more variability in example 1 even though the S.D. is smaller. These examples could be a comparison of sales of expensive items with the sales of less expensive items.

The data in Tables 3–5 and 3–6 were comparable sets—the same salespersons selling comparable units over comparable sales periods. The number of salespersons could have been different but the units and periods should be alike in order to compare the mean and standard deviation of the two sets of data. You know your salespersons are about equal when the standard deviation is small.

Assume now that we want to see if the number of contacts made is out of line with the number of sales made. We expect that one contact will result in more than one unit sold. We cannot compare two sets of data, using the mean and S.D., if one set is expected to have larger values than the other. For example, mean of number of contacts may be 20, whereas mean of units sold may be 300; S.D. of number of contacts may be 5, whereas S.D. of units sold may be 50. We can, however, compare last sales period with this sales period using the same statistics, as we did in Tables 3–5 and 3–6, or we can compare sales in one region with sales in another region.

Table 3–7 shows the number of contacts made by the seventy salespersons during the same sales period. Skewness and kurtosis are negligible and the mean, median, and mode are very close. These combined conditions denote a fairly normal distribution. As stated previously, we cannot compare this standard deviation of 5 with that of 50 in Table 3–5. It is also reasonable that smaller numbers would force a smaller standard deviation. A standard deviation of 50 would be impossible when the maximum value is 35 with no negative numbers in the set. The size of the standard deviation is relative to the size of the original values and the range of the set of data. We have

Number of Contacts	Number of Salespersons	Relative Frequency Percent	Cumulative Frequency Percent
13	2	2.9	2.9
14	4	5.7	8.6
16	2	2.9	11.4
17	6	8.6	20.0
18	2	2.9	22.9
19	6	8.6	31.4
20	8	11.4	42.9
21	5	7.1	50.0
22	9	12.9	62.9
23	4	5.7	68.6
24	3	4.3	72.9
25	5	7.1	80.0
26	5	7.1	87.1
27	2	2.9	90.0
30	3	4.3	94.3
31	1	1.4	95.7
35	3	4.3	100.0
Total	70		

Minimum	13	Median 21.5
Maximum	35	Mode 22
Range	22	Mean 22 (rounded)
Skewness	0.65	S.D. 5
Kurtosis	0.58	(A span of 5 is 1 S.D.)
Std. Error	0.5	C.V.% 23
95% C.I.	21-23	

Table 3-7 Frequency Table—Sales Contacts

to look at the coefficient of variation and the standard deviation expressed as a z-score. To compare this set of data in Table 3-7 with that in Table 3-5 we use the value of the coefficient of variation, C.V. = 23 percent. C.V. for Table 3-5, sales, was 15 percent. Number of contacts showed more variability than number of sales. The salespeople were not as consistent in contacts made.

Now let's investigate the relative standing of one hypothetical salesperson. We will assume that he or she made 19 contacts and 360 sales. Looking back at the interpretation of Table 3–5 we found 360 sales to be within +1 S.D. (3.b.), a little above average. This also was shown in Figure 3–6 as the 322 to 372 bar. The 19 contacts entry falls in the area of −1 S.D. The mean of 22 minus the 1 standard deviation of 5 includes 17 to 22 contacts, meaning 19 contacts was a little below average. For exact figures or z-scores on the data for this salesperson who made 360 sales and 19 contacts, we can use the formula given previously in this chapter:

$$z = \frac{X - \text{mean}}{\text{S.D.}}$$

for sales:

$$z = \frac{360 - 322}{50}$$

$$z = \frac{38}{50}$$

$$z = .76$$

for contacts:

$$z = \frac{19 - 22}{5}$$

$$z = \frac{-3}{5}$$

$$z = -.6$$

The z-score for contacts for this person is shown in Figure 3–8 under the value 19, along with the other data from Table 3–7. Each dot represents one person who made the number of contacts for the values shown (or implied) below the line. The lower row of figures from −3 to +3 is the standard deviation expressed as a z-score. Figure 3–8 also illustrates a fairly normal distribution—more concentration of data near the mean and less out toward both ends. This fact was noted for Table 3–7 with the mean, median, and mode close, and with negligible skewness. This salesperson was a little above average in sales (+.76) and a little below average in contacts (−.6). What could more contacts bring for this salesperson? Maybe this example warrants no action; how about the possibility of a salesperson who is +2 S.D. in contacts and −1 S.D. in sales? Of course, with this formula, the computer can produce this type of information on each salesperson. We do not have to do it manually as in the example. After the computer has done the calculations, we can also receive a list of those salespeople who do not meet the criteria we think they should.

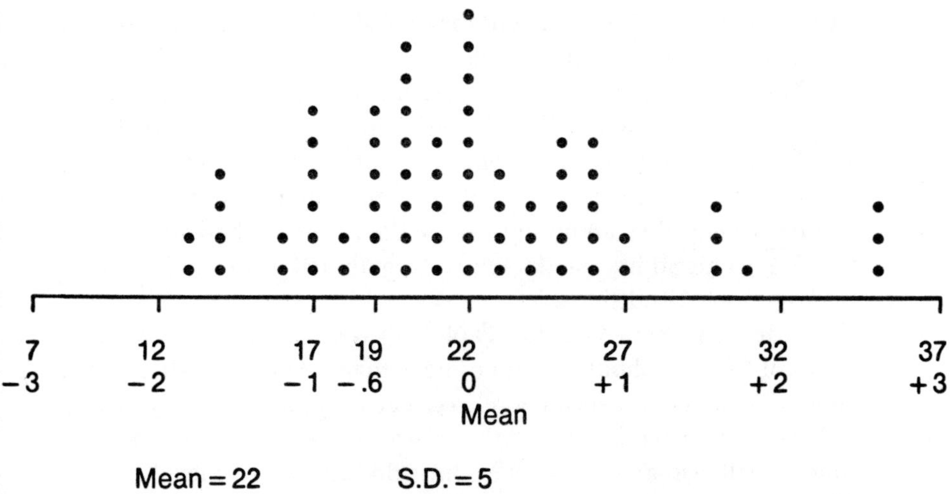

Figure 3-8 Modified Frequency Distribution Illustrating Standard Deviation, z-score for Sales Contacts

The statistics we receive from a computer based on the sample data we have given it to compute allow us to make a very close guess about what would occur if we could collect all the data possible. Much data is finite—internal personnel data, for example. When all the data is accessible, it should all be used. Computers allow us to do this easily, thereby giving us more accurate results on which to make decisions. Other data is infinite or inaccessible and has to be interpreted by taking into consideration an estimate. Although as much data as is available should be used, the addition of more data that is related to that which you have should not change the results appreciably. How much the results might change is determined by an estimated error factor. This error factor, called a **standard error**, is based on the variation (standard deviation) and the number in the group. As the number of cases increases, the value of the standard error decreases (more cases, less chance for error); also, as the standard deviation decreases, the value of the standard error decreases (less variation, less chance for error). The standard error is similar in use to the standard deviation except that the 95 percent level is more often used. Remember, approximately 95 percent of data should fall within +2 or −2 standard deviations; i.e., the calculation of +2 or −2 standard deviations would tell us within what range 95 percent of the data should fall. The calculation of +2 or −2 standard errors will tell us within what range the mean of related data should fall about 95 percent of the time. This +2 or −2 standard errors is called the 95% **confidence interval**. If you see *.95 C.I.* or *95% C.I.* on computer output, the interpretation is: 95 percent of the time, or with 95 percent confidence, the statistics will be similar after addition of more data. (more about this in section 5.4.) This additional data must be gathered from similar groups under similar conditions.

The 95 percent C.I. for Table 3–5 was 310 to 334. Using the mean of 322, we should have no more than 12 points' difference in either direction in the mean of related data (or future data). The standard error produced by the computer was 6, and +2 times 6 and −2 times 6 will also give us the 12 points difference in either direction. The 95 percent C.I. for Table 3–7 was 21 to 23. Using the mean of 22, we should have no more than one point difference in either direction. The standard error was .5; similarly, +2 times .5 and −2 times .5 will give us the one point difference in either direction.

Data, as a whole, is commonly described by the use of averages and standard deviations and to a lesser degree by skewness and kurtosis. The relative standing of specific values within that set of data is described by the z-score. The computer provides the numbers; you must interpret them and make the decision about the reasonableness of the numbers and the need to know why if the numbers are not what you expected.

4

Showing Relationships

1. DETERMINING RELATIONSHIPS

Common sense analysis—this is where the manager may have an advantage over the statistician. The manager is closer to the situation, has a better feel for what characteristics or factors may be related, and is in a better position to notice factors occurring together. A relationship is a connection of some kind. Finding two characteristics often occurring together leads us to believe that there is some connection between them. A causes B, or maybe B causes A, or maybe a third factor C causes both A and B. We want to find the cause—of either good results or bad results. We want to determine if our chosen course of action will produce the desired results; we want to pinpoint a trend; we want to support our educated guesses with facts. We need the common sense analysis mentioned above; figures cannot always be quoted as facts and, at worst, they can be misleading. For example: In a small town election, the sheriff won on a promise to cut crime in the streets in *half*. After winning, the new sheriff made good on the promise—by constructing *twice* as many streets. The numbers add up but the result is ridiculous.

If we think, using our common sense or educated guesses, that there is a relationship between two things, then we need some way to compare them, some way to back up our ideas with facts. People often ask, "Is there any correlation between. . .?" The correlation coefficient, or simply, **correlation** provides a way of comparing and measuring the relationship of two qualities or characteristics of a set of data. These qualities are expressed as numbers obtained from some kind of measurement. (The comparison of more than two qualities, using correlation, will not be covered in this book.) This relationship measurement is given in one number that ranges from −1 to +1, with 0 being no relationship between two qualities. The negative in this case refers to the type of relationship and will be covered in the next section. The closer to −1, or to +1, the stronger the relationship, and the closer to 0 the less relationship, or correlation.

Educators have always compared data on aptitude tests, IQ tests, achievement tests, and grade point averages. They want to predict educational success from tests or to judge the validity of their tests from educational success; they stress the correlation between income and education. Personnel managers use employment tests, a known relationship of certain qualities, to predict on-the-job performance. Geneticists have worked with correlating physical characteristics, as have anthropologists with man's development and customs. Medical researchers have noticed that certain conditions seem to be related to certain diseases. Agronomists compare weather or amount of fertilizer with crop growth.

For staffing or production quotas, we may want to refine our prediction of the number of customers we can expect by time of day, or by day of week, or by day of month. We may want to predict how much profit we need to hire additional staff. If we keep a record of age of equipment and repair costs, maybe we can make a better judgment about when to replace the equipment. We may want to compare the amount spent on advertising with sales during that period, or the effect of a higher price on sales; or we may want to compare salaries to productivity. We may want to determine if employee absenteeism is related to job type.

As we look more closely at correlation, keep in mind that *compare means pair.* Possible pairs in correlation include: the competition or comparison of people with themselves (test scores and grade point average); the comparison of a machine's age with the repair costs of the same machine; the comparison of advertising costs for one month with sales for the same month; the comparison of preconditioning skills with postconditioning skills for a person.

2. USING GRAPHS TO HIGHLIGHT RELATIONSHIPS

After we are sure a relationship exists, we can produce a simple formula. For example, a person earning $10.00 per hour for eight hours earns \$80.00. (We use \$10.00 for ease of computation.) For any number of hours we have:

salary = \$10.00 times the number of hours worked, or

$$S = 10n.$$

There is a relationship between S and n. As n increases by 1, S increases by 10. Figure 4–1 shows this paired relationship on a graph. It is called a **scatter diagram.** The salary is on the vertical line—the y-axis. You will recall this terminology from high-school algebra. Salary is dependent upon number of hours worked. Therefore, it is called the dependent variable. Number of hours worked becomes the independent variable and is on the x-axis of the graph. We can predict salary from hours worked. The rule is: Variables (section 1.2) we try to predict are called **dependent variables** and are on the y-axis; variables we are basing our prediction on are called **independent variables** and are on the x-axis. Computer statistics may not label these *x* and *y*, but the computer has to be told which is dependent and which is independent. Notice that the dots (pair of values) form a straight line on the graph in Figure 4–1 and this imaginary line slopes upward from left to right. As x increases, y increases. These conditions are necessary for a correlation of +1.

Next, let's assume this person is not on hourly wages. The person receives \$80.00 a day whether he or she works one hour or twelve

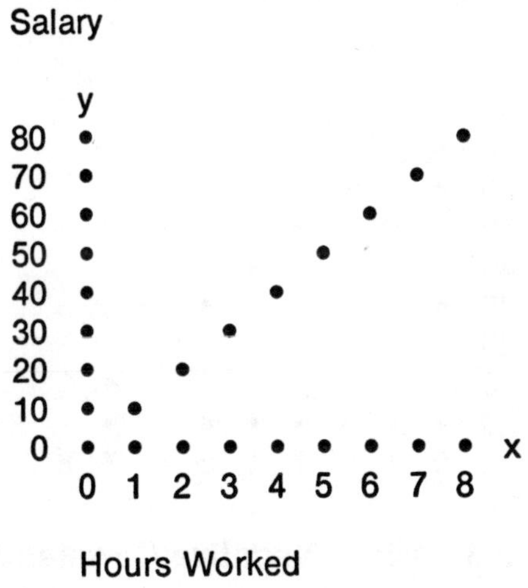

Figure 4–1 Perfect Positive Correlation (+1)

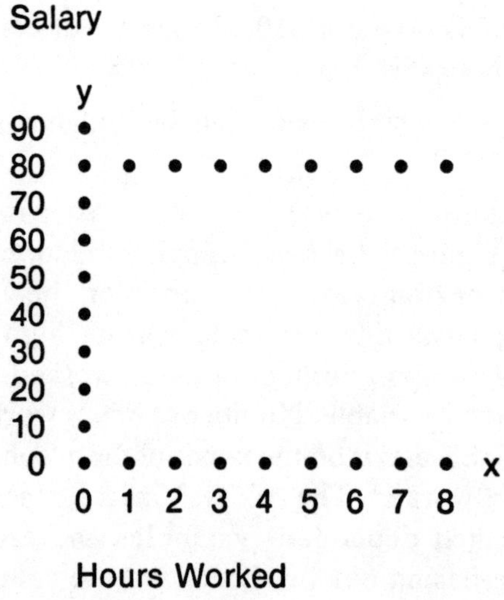

Figure 4–2 No Correlation (0)

hours. The graph would look like Figure 4–2. There is no relationship between salary and hours worked. The imaginary line is parallel to the x-axis.

So far, we have looked at a correlation of +1 and a correlation of 0. You've probably guessed that a correlation of −1 would be the opposite of +1. Figure 4–3 shows the correlation of −1 in graph form.

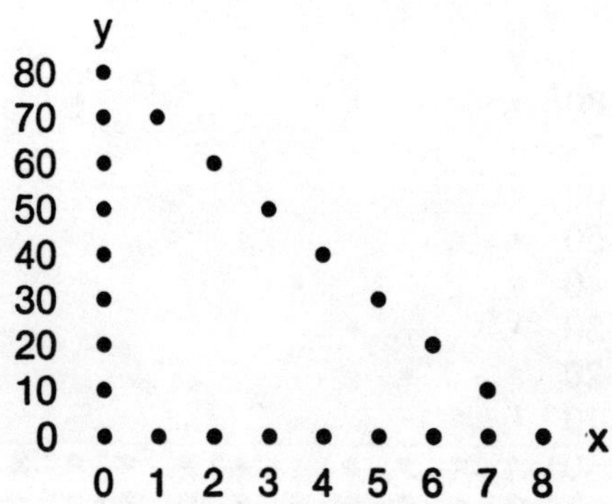

Figure 4–3 Perfect Negative Correlation (−1)

As x increases, y decreases, but there is still a perfect relationship with the imaginary line now sloping downward from left to right. A correlation of -1 is no weaker than a correlation of $+1$; it is simply an inverse relationship.

The preceding figures in this chapter were for illustration and, in our comparisons with real data, would occur only under special or controlled circumstances. Let's get back to the real world and look at the types of graphs we are most likely to get in comparing one quality or variable with another. As stated previously, we are considering simple correlation in which we have a pair of numbers for a person, or for a machine, or for a treatment, etc. These pairs of numbers are plotted on a scatter diagram, and if they appear to be close to a straight line we can assume there is a relationship between them—what is called a linear (or straight-line) relationship. (Other nonlinear relationships will not be covered in this book). The imaginary straight line that approximates this relationship is called a **regression** line and can be used as a prediction line (see section 4.4). Figures 4–4 through 4–8 show examples of the general appearance of the graphs under different conditions. Figures 4–4 and 4–5 show data points that are very close to a straight line; in Figures 4–6 and 4–7, the data points

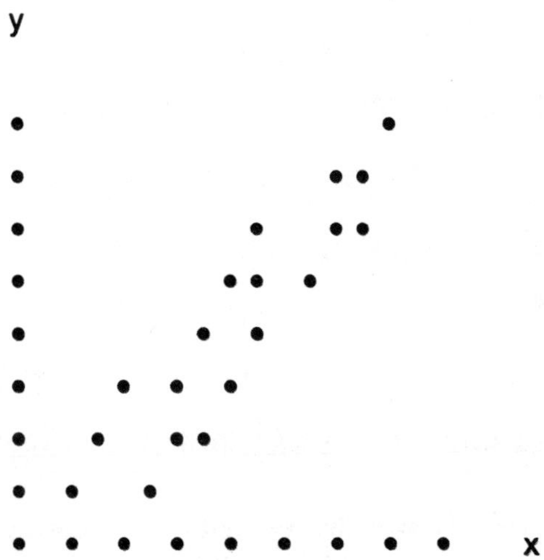

Figure 4-4 High Positive Correlation—Close to +1

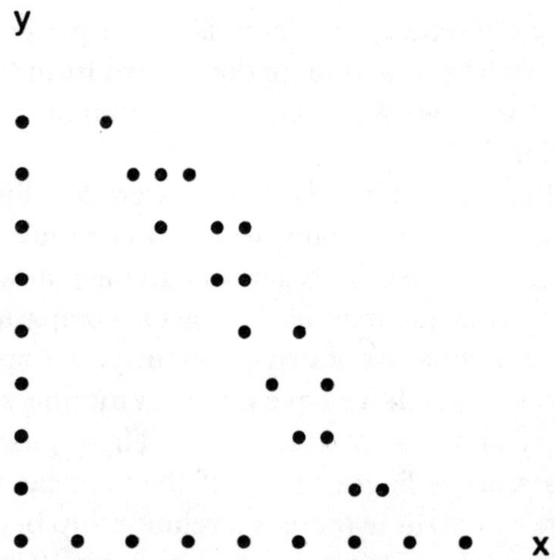

Figure 4-5 High Negative Correlation—Close to −1

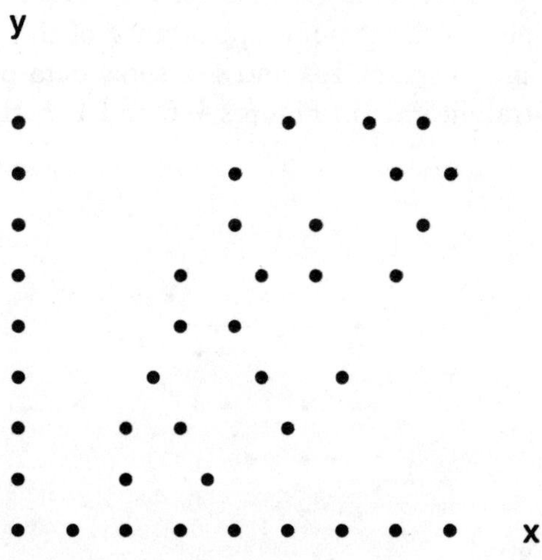

Figure 4-6 Low Positive Correlation—Close to .5

give the appearance of a straight line, but are more scattered or spread out; and the data points in Figure 4-8 definitely do not resemble a straight line.

The scatter diagram is all that is needed if you want to show only correlation and are not interested in the degree of correlation. Like

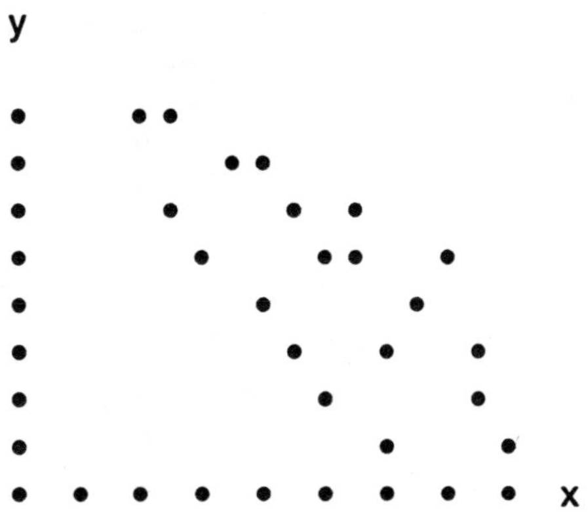

Figure 4-7 Low Negative Correlation—Close to −.5

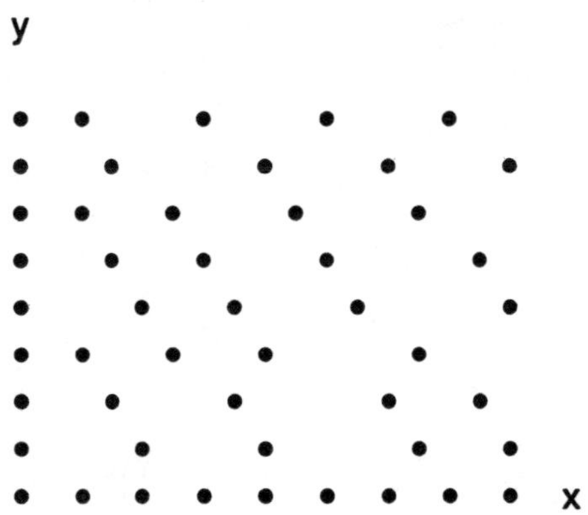

Figure 4-8 No Correlation—Close to 0

the bar graph (section 2.2), it gives a picture of what the data looks like. In addition, you can see the correlation, and if it is positive or negative. Does it approximate a straight line sloping upward, or a straight line sloping downward? If not, there is no correlation of the simple linear type. The scatter diagram should have the line for the x-axis and the line for the y-axis approximately the same length so that one variable is not shown to be extreme in relation to the other variable. The result of a distortion (or a trick) of this type can be seen in Figure 4–9. The computer should avoid incorrect representation

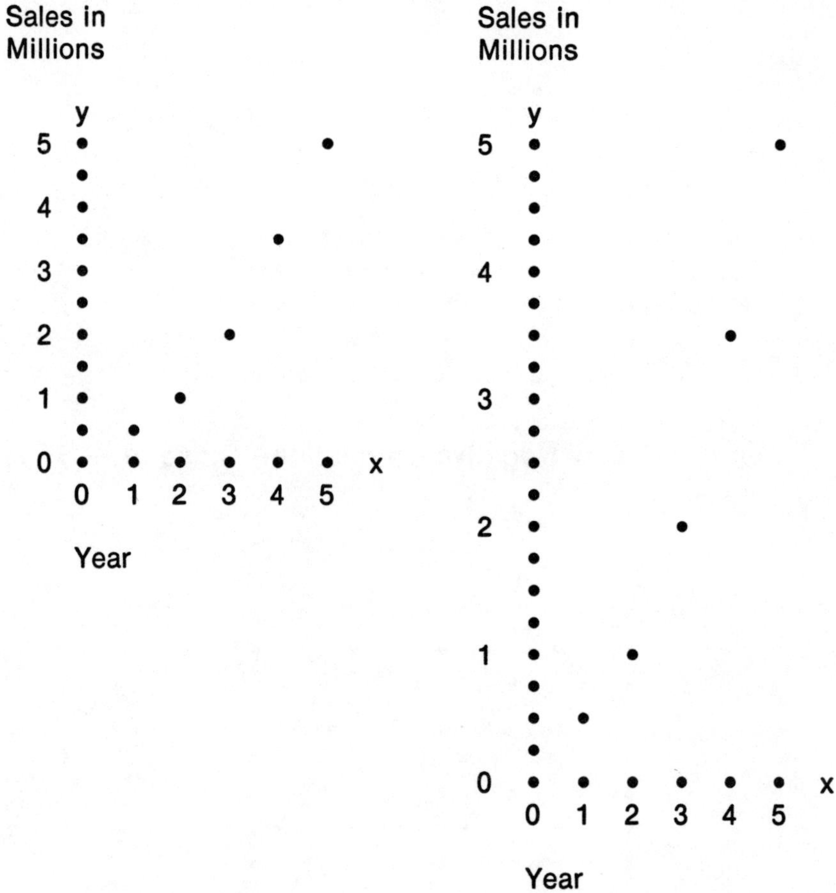

Left: Correct representation—x and y axis approximately the same length
Right: Incorrect representation—sales trend shown to be extreme from year to year

Figure 4-9 Distortion of Graph

by a method of division and scaling—using, for each variable, the lowest value and the highest value that appear in each set of data. For example, values for x are: 15, 20, 22, 31, 40, 47, 49, 55, 58, 65. The lowest value is 15 and the highest value is 65, a range of

50. If there were space for eleven designations on the x-axis, it could look like this.

```
•    •    •    •    •    •    •    •    •    •    •    x
15   20   25   30   35   40   45   50   55   60   65
```

One way the computer can obtain this result is by dividing the range, 50 in this case, by the number of designations minus one, or 11 − 1; 50 divided by 10 gives the unit of 5 shown for the x-axis. The same method would be used for the y-axis, then the pairs of data points can be plotted at the appropriate points for each pair of x and y values.

If more precision is necessary, you may request the scatter diagram and the correlation statistics or only the statistics without the scatter diagram. You will probably find a limitation on the number of pairs the computer will plot, at least on the number of identical pairs, if not on the total number. The statistics will include a correlation coefficient between −1 and +1 usually denoted by the letter, *r*. This figure tells us how close all the data points come to being on a straight line when graphed (see section 4.4), after we are sure the data points are approximately linear on a scatter diagram.

3. ANALYZING THE EXTENT OF RELATIONSHIP

The next example has been manufactured to show a high positive correlation. We will assume there is a correlation between the amount spent on advertising and the amount of income, with income dependent upon advertising. Table 4–1 shows data from the last fifteen months, not necessarily in order by month. The computer has arranged it in ascending order by advertising costs. Printing of the month and year could have been requested for our additional information but only the pair of dollar figures from each month is used in the correlation. Figure 4–10 shows the data as a scatter diagram. The number 2 within the graph denotes two months in which $2,900 was spent on advertising and $17,000 was reported as income. We can see by the near straight line on the graph that there is a high correlation. This may be all we need to know—a verification that advertising money is not being wasted—and only you would know if there are some factors contributing to the income other than advertising. The computer statistics is an additional tool in facilitating the record keep-

Advertising Costs	Income
2100	15000
2300	15000
2400	16000
2900	17000
2900	17000
3000	18000
3200	19000
3500	21000
3700	21000
3800	22000
4000	23000
4200	24000
4300	25000
4600	27000
5000	30000

Table 4-1 Advertising

ing of the advertising manager. More detail in the company's computer records (see Chapter 6) will allow a breakdown by type of advertising—TV, radio, magazines—or special advertising campaigns.

Suppose we think there should be a positive correlation between contacts made and units sold. We expect that more contacts will result in more units sold. The first step is to have the computer pair the contacts made with the units sold for each salesperson and, if wanted, produce a scatter diagram. We will use the data from Tables 3-5 and 3-7. As we believe that the units sold should be dependent upon the contacts made, the units sold will be the dependent variable on the y-axis and the contacts made will be the independent variable on the x-axis, as shown in Figure 4-11. By no stretch of the imagination can we approximate a straight line from this graph. Additionally, the data for one salesperson shows 21 contacts and 490 units sold, denoted by point A on the graph, while another salesperson made 35 contacts and sold only 240 units, denoted by point B on the graph. The computer reported a correlation coefficient, $r = -.13$, for this data. This is very close to 0, or no correlation. This should prompt some investigation by the supervisor about the lack of correlation. Part of

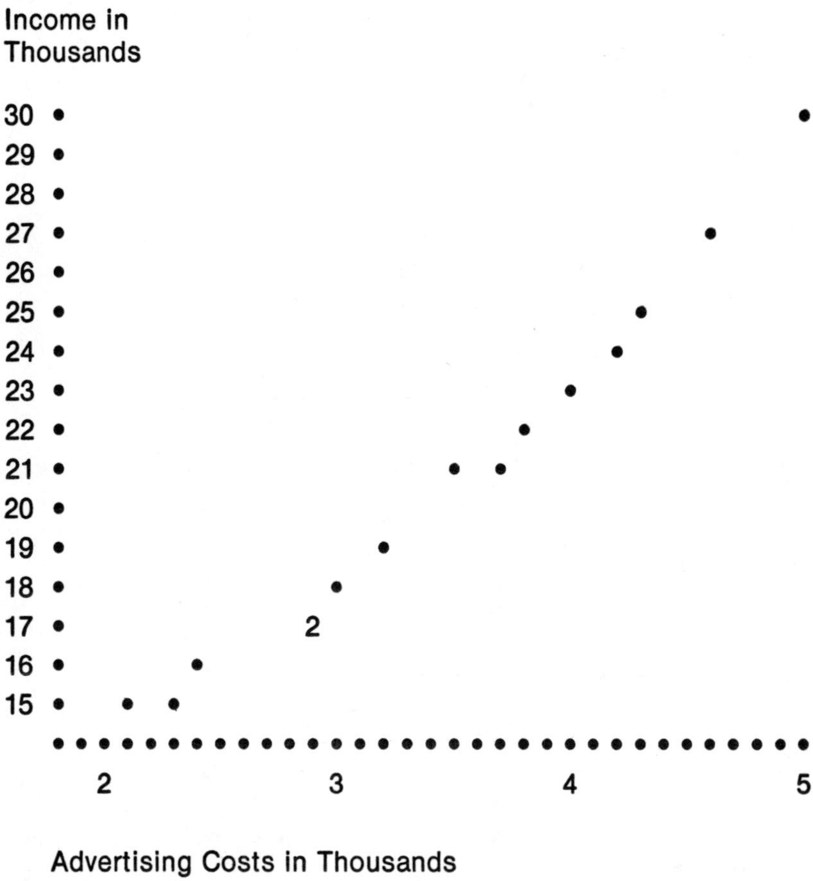

Figure 4-10 Advertising: r = .99 (rounded)—Scatter Diagram

this investigation should include the addition of more data over a longer period of time; the separation of experienced salespeople from those inexperienced; and separation of data by day of week or by day of month. This is easily done by computer if your computer files are complete (see section 6.3). Special travel allowances or possibly inequitable sales regions could be a consideration.

In our computer inventory records, part of each record contains the number of units on hand and the cost per unit. Although we are protected from overstocking or understocking by a reorder point also on the record, we want to see how cost compared to units on hand.

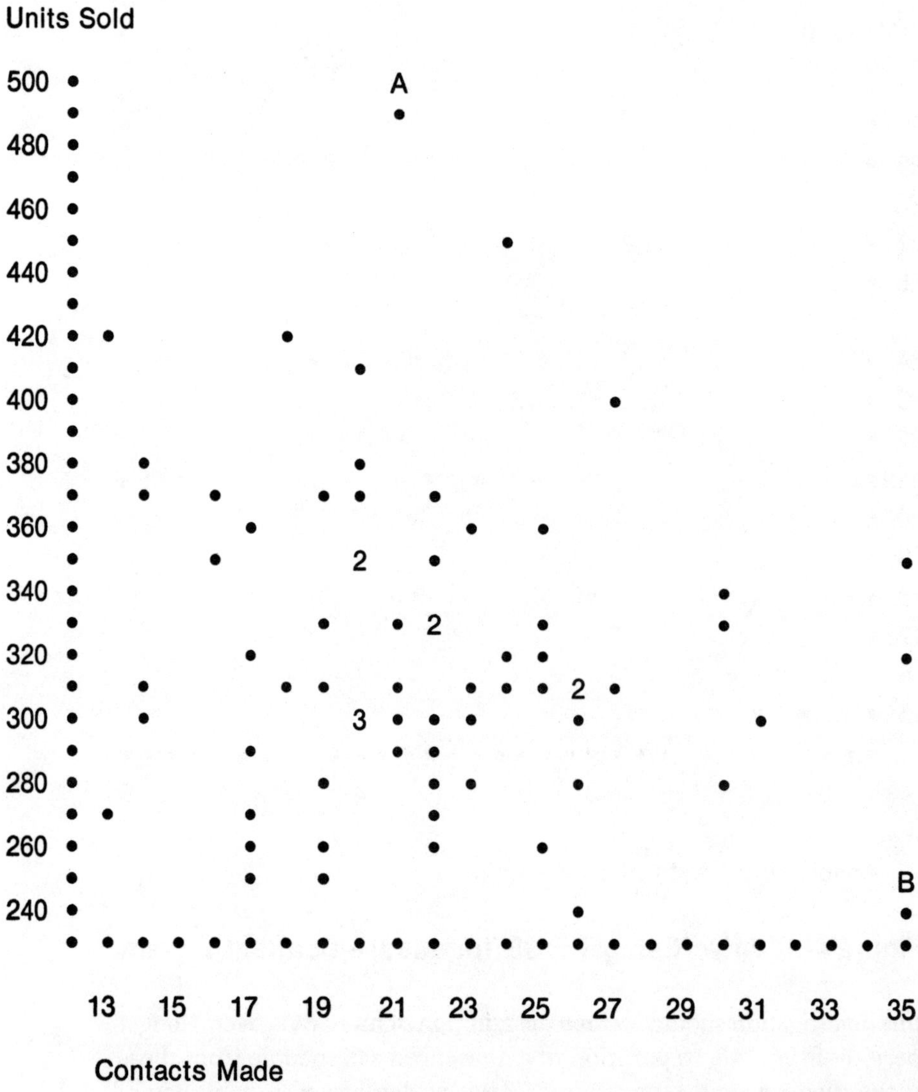

Figure 4-11 Sales Correlation: r = −.13—Scatter Diagram

We also want an easy way to check the appropriateness of our inventory. As in previous examples, the data is selected for illustration, so we will not get bogged down in detail. Table 4-2 shows that portion of the inventory records being compared. A computer statistics

Cost per Unit	Units on Hand
900	12
850	17
800	20
750	21
700	25
650	30
600	33
550	40
500	40
450	50
400	43
350	46
300	59
250	57
200	70
150	80
100	90
50	100

Table 4-2 Inventory by Cost

package should also provide the facility for producing total costs and other vital information for inventory.

Figure 4–12 is a scatter diagram of this third important type of relationship—the high negative correlation, with $r = -.97$. As the cost of the unit increases, the number of units on hand decreases, and we can approximate a straight line sloping downward.

A familiar example of negative correlation is rising interest rates producing a decline in stock prices, gold, silver, bonds.

4. USING A RELATIONSHIP TO PREDICT

In section 4.2, a simple equation for salary ($S = 10n$) was given. Ignoring all other factors related to salary, we determined or predicted salary by substituting number of hours worked in the equation. In many correlation studies we want to see either if two qualities are related (always seem to occur together) or if they are definitely not related; we are not interested in predicting what will happen to one

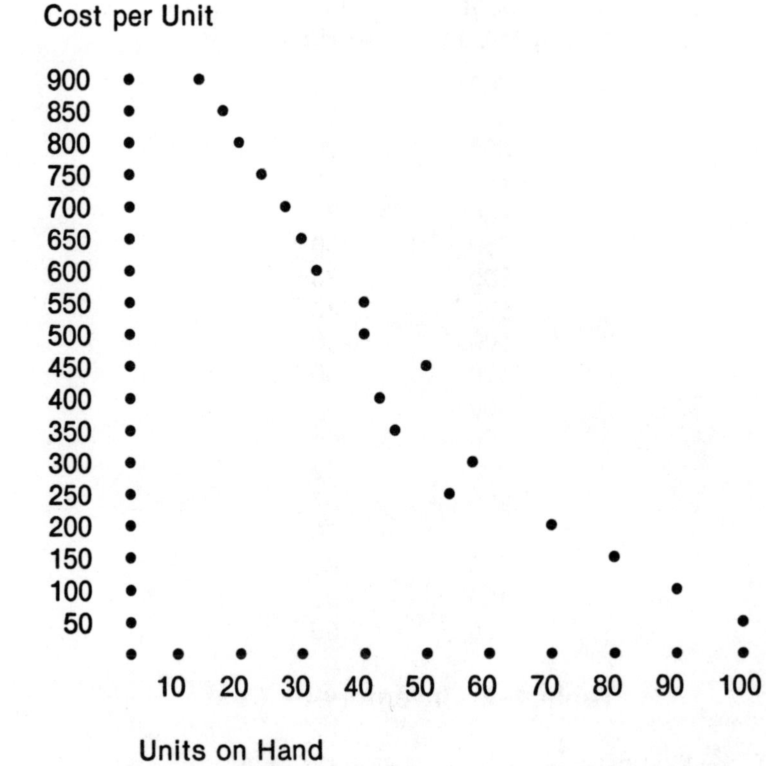

Figure 4-12 Inventory by Cost: r = −.97 (rounded)—Scatter Diagram

variable if we change the other variable. In other studies this is the main purpose—predicting college success from test scores; age at which equipment should be replaced; dosage amount to achieve certain results; or income needed for certain conditions.

Also in section 4.2, we referred to a regression line. We could look at the scatter diagram for hours worked, find the number 3, for instance, and then determine that this was located opposite $30.00. Of course, the easier method was to substitute 3 in our simple equation. The regression equation is similar to the salary equation we illustrated, but it takes into consideration the value of one variable when the other is 0. This can be thought of in terms of paying insurance premiums for employees when they are not working. We have to add a constant value to the equation so that it becomes S = i + 10n. Now let's rewrite the equation we used for illustration to the form usually seen in statistics: y = a + bx, where *y* is the dependent variable we are predicting, *x* is the independent variable we are basing our prediction on, *b* represents the change in y (10 in our salary example) for every unit change in x, and *a* is the constant to be added. In computer statistics you may see *b* denoted as slope and *a* denoted as y-intercept, or simply, intercept. (In high school algebra, you probably saw this equation as y = mx + b, where slope m is multiplied by x and b is the intercept).

Before using this equation as a predictor, we must determine by a scatter diagram that there is a linear relationship between the two variables, x and y. We did this for the advertising-income example, Figure 4–10, and obtained r = .99. Among the statistics produced by the computer were the values of a and b for the regression equation: slope b = 5 (rounded) and intercept a = 2900 (rounded). The slope b of 5 can be interpreted thus: for each increase of $1.00 spent on advertising, income should increase by $5.00, or for each change in x of 1, y will change by 5. Intercept a of 2900 shows what income could be expected if nothing is spent on advertising, or the value of y when x is 0, so for this particular example, the equation,

$$y = a + bx \text{ becomes}$$
$$y = 2900 + 5x$$

We now have the regression equation produced by the computer. All we have to do is substitute a value for x, advertising, and predict the value of y, income. This can be done manually or by computer. Let's make it easy and assume we want to predict income if $10,000 is spent on advertising. We have

$$y = a + bx$$
$$y = 2900 + 5(10{,}000)$$
$$y = 2900 + 50{,}000$$
$$y = \$52{,}900.$$

As in all correlations and predictions, we have to use our common sense about the reasonableness of these figures, and we have to make the assumption that all other factors influencing income will remain the same in the future. If other factors change, then our prediction equation is invalid and we have to repeat the study. *Predictions of the future are based on patterns of the past.* Additional factors that influence these patterns will have to be considered as they change, and some linear patterns of the past may not stay linear in the future. When you are ready for more advanced statistics you will want to investigate nonlinear patterns in other statistics books.

For the inventory example, Figure 4–12, the computer produced the values of slope b = −10 and intercept a = 941. If we wish to use the same relationship of cost and units on hand, then for $20.00 items (x), we would stock 741 units (y). This was determined by substituting these values: b = −10, a = 941, and x = 20, thus,

$$
\begin{aligned}
y &= a + bx \\
y &= 941 - 10(20) \\
y &= 941 - 200 \\
y &= 741.
\end{aligned}
$$

This inventory example was based only on the cost of the item. Other considerations would include the storage space required and the backlog; and maybe we should compare stock on hand with stock sold per month.

Also in the inventory example, we saw that a negative slope (b = −10) indicates that the straight line will slope downward, and we will subtract in the regression equation. Correlation, r, will have the same positive or negative sign as the slope. If slope b is 0, our equation,

$$
\begin{aligned}
y &= a + bx \text{ becomes:} \\
y &= a + 0(x), \text{ or} \\
y &= a.
\end{aligned}
$$

This was true in Figure 4–2 (shown here as Figure 4–13), showing no correlation. The imaginary straight line is parallel to the x-axis. It follows that if b = 0, then r = 0. In Figure 4–13, using y = a, y = 80 for every value of x; there is 0, or no, slope and no relationship between x and y (hours worked and salary). We do not have to know the value of x to predict y; that is, we cannot use this as a prediction equation because knowing the value of x is no help in predicting y.

As stated previously, b represents the change in y for every unit change in x. Consider the numbers for x and y shown in Figure 4–14. Slope, b, can be found easily for this example; the change in y is 2 units as we look down the y column, and the graph in Figure 4–14 shows that all data points are on a straight line. Everything and everyone does not fit this neat pattern (fortunately). For the data that

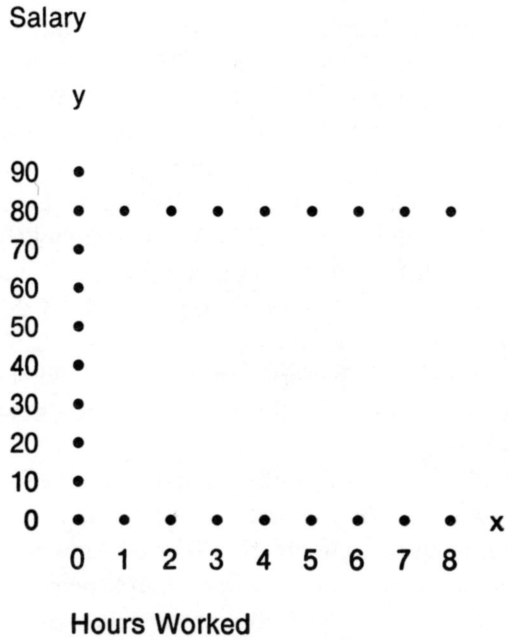

Figure 4-13 No Correlation, r = o
No Slope, b = o

x	y
0	0
1	2
2	4
3	6
4	8

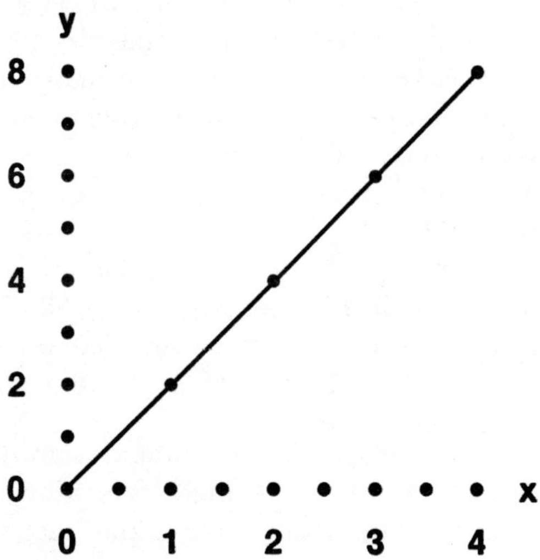

Figure 4-14 Illustration of Slope

does not fit a pattern, there will be some error in our predictions or comparisons. We need to know how much error or how closely the data does, or does not, fit a pattern. Let's add a couple of data points, 5 and 6 for x, to the graph in Figure 4–14. This is shown in Figure 4–15. The vertical lines represent the differences between the actual data points (F and G) and the predicted data points (on the line) that we would have had if the additional data points had followed the same pattern as A through E. These differences, +2 for F and −3 for G, contribute to errors in estimate (see section 4.5). The slope of the straight line we had with the first four data points will change as data points that do not fit this pattern exactly are added. How much it changes will be determined by the distances of the new data points from the line.

The several versions of mathematical formulas for finding the regression equation will not concern us in detail. One is a method you may have encountered in algebra—the *least-squares* method. The name suggests minimum and squaring—more specifically, the sums of the squares of the distances between the actual data points, such as the two data points, F and G, in Figure 4–15, and the corresponding data points on the regression line. The distance between point F and the line and the distance between point G and the line are squared and added together. Squaring avoids the possible problem of getting a result of 0 when some data points are below the line and some data points are above the line. Many statistics are eventually obtained by squaring some kind of measure, in many cases to avoid negatives cancelling out positives when addition or subtraction is also used in the statistical formula. These sums of differences are used to produce an error factor, which will be explained in section 4.5. Of course, adjusting the line for these two data points will result in having A through E not exactly on the line, but we do not continually change the line as implied in the illustration. All of the data is considered at one time, and the computer finds the best-fitting line. The result will be a regression line in which the distance of each data point from this line is minimized.

Correlation, r, will be a number showing how closely the data fits this pattern or line. A high correlation, close to +1 or to −1, means most of the data is close to this line; a low correlation, closer to 0, means the data is more scattered over the graph. The mathematical formula for r considers the distance of each x value from

x	y
0	0
1	2
2	4
3	6
4	8
5	12
6	9

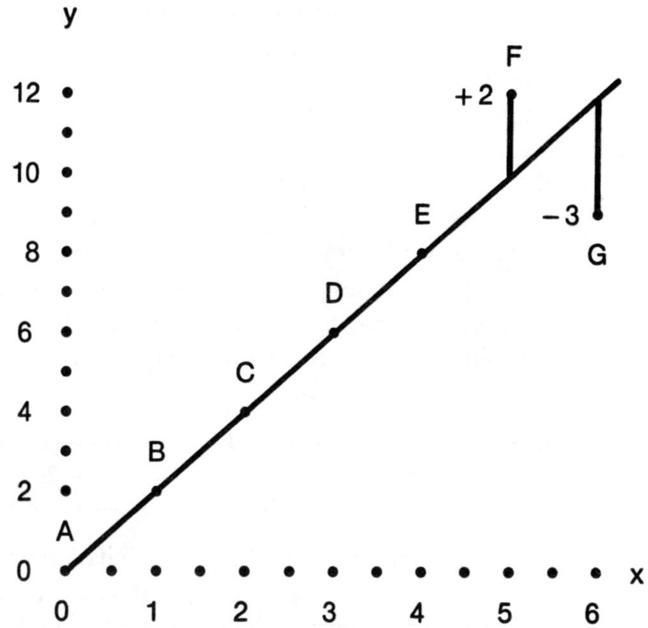

Figure 4–15 Finding the Regression Equation

the mean (section 3.1) of x values and the distance of each y value from the mean of y values. Since the mean is used, we know the number of cases has to be used.

5. RISK OF BEING WRONG

If you became ill after the cocktail party, did you blame it on the cheese crackers you ate? Correlation may be the most difficult statistic to interpret. There are so many factors influencing the relationship of two variables, especially the human factor. It is difficult, if not impossible, to isolate only the two qualities we are comparing. The best we can do is ensure that the comparison makes sense, compare data that is comparable, use as much data as possible, and not jump to conclusions about the results. We have to use our best judgment about the strength of the correlation by taking into consideration the factors we cannot isolate. For example, the correlation between aptitude tests or high school grade-point averages with college grade-point averages for college freshmen may be lower than the correlation determined on the same group as sophomores or juniors. The newness of the situation is no longer a big factor, and to a certain extent, age, interests, and other factors have changed.

In general, a correlation of r = +.80 or −.80 or better is considered good, r = +.30 or −.30 or less is considered poor or no correlation. Those in between are a risk and should be interpreted according to the importance you place on the study, the precision necessary, your judgment of the cost of making a wrong decision, and in some cases, the cost or time involved in collecting additional data. This range of correlation from none (or insignificant) to high is shown with the modified number line.

```
I----I----I----I----I----I----I----I----I----I----I----I----I
-1  -.80 -.60 -.40 -.20                   +.20 +.40 +.60 +.80  +1
High-I---Caution--I----None-----I--Caution---I-High
```

To help with interpretation of correlation and precision, the computer produces two other statistics that are worth attention. One is **r squared** (coefficient of determination) which indicates how much of the relationship is due to the factors we are comparing. One way this is obtained is the way the name implies—square the correlation, r times r. This will be addressed later in this section.

The other statistic is useful with the regression, or prediction, equation. If all the data points on the scatter diagram fall on a straight line, there is no error involved in predicting y by substituting different values for x. Realistically, we cannot be 100 percent right in our predictions because we never have all the pairs lying on a straight line, and we do not have an infinite amount of data. We can only say, "chances are y will be approximately. . . ." The **standard error of estimate** provides a number that shows how much variability can be expected in the prediction of y. The standard error of estimate (S.E.E.) is a measure of variability around the regression line. (The standard deviation was a measure of variability around the mean.) S.E.E. is produced by the computer, using a formula that considers the differences mentioned in section 4.4 (Figure 4–15) and how much data is used; thus, the more data, the less chance of error in predicting. While the regression line minimizes the differences, S.E.E. gives us a number showing how much difference there is—how accurate our predictions are. If there were only two cases (two data points) there would be no error; the straight line would pass through these two points. However, using only two data points would give us no basis for estimating universal patterns. The word "standard" refers to the standard deviation of future data that should follow the same pattern. All of us are familiar with standardized tests—those that provide a standard by which to judge. The same or very similar tests have been given and regiven; the statistics (mean, standard deviation, etc.) are known from past applications and the test has been shown to measure what it purports to measure (high correlation between x and y). Thus we have a standard by which to compare present data, or to predict future data, based upon past comparable data. This is the purpose of the standard error of estimate for regression.

R squared measures the explained variation—variation in y that can be attributed to the variation in x. The unexplained variation is chance or other factors, and is calculated using the same differences shown by example in Figure 4–15. R squared and unexplained variation together account for 100 percent of the variation. Changing 100 percent to its decimal equivalent, 1.00, we have

$$\text{r squared} + \text{unexplained variation} = 1.$$

Thus neither can be larger than 1. Both vary from 0 to 1. R squared = 1 means all of the variation in y is explained by the variation in x and S.E.E. would be 0; all of the data points would fall on a straight line. R squared = 0 means none of the variation in y is explained by the variation in x. Correlation, r, would also be 0, since 0 squared is 0. The less correlation between variables, the more error in predicting. Other than squaring r, r squared can be obtained by subtracting the calculated unexplained variation from 1. In other words, there is a direct relationship between correlation and amount of error, both measuring the spread of data.

For the income-advertising example (Figure 4–10), r squared was .97 meaning it is likely that 97 percent of the variation in income can be attributed to the advertising and 3 percent, the unexplained variation, can be attributed to other factors (.97 + .03 = 1), or almost all of our income is dependent upon our advertising. Considering the income-advertising example again, the standard error of estimate was \$762, and y, or income was estimated to be \$52,900. Relative to the prediction of \$52,900, \$762 is a small error. These figures indicate that with the expenditure of $10,000 for advertising, income should be $52,900, give or take $762 (or, in round numbers, give or take a thousand). An error of perhaps $4,000 may be too large an error when considering future data of the same type with approximately the same number of cases. If this is the case, we cannot use this as a predictor because the error would be too large. There is an advantage in the advertising study that many correlation studies do not have. We can check the accuracy of our predictions after spending x amount of dollars for advertising and check how closely income, y, comes to the prediction we had made.

6. SUMMARY

Let's consolidate the main points covered in this chapter with one more example:

1. Correlation: r, a number between -1 and $+1$ showing the relationship between two sets of continuous data items.
2. Explained variation: r squared, a number showing what percent of the variation in y is due to the variation in x.
3. Regression equation: $y = a + bx$, a formula for predicting y (dependent variable) for any value of x (independent variable), if x and y are known to be linearly related.
4. Standard error of estimate: a number that shows how much variability can be expected in the prediction of y.

The following example uses an aptitude test to predict the score a person should make on an achievement test. This combination, generally, should assure success in a chosen subject or job. Past successes, and failures, are paired by person in Table 4–3 and again in the scatter diagram, Figure 4–16. As we are trying to predict the achievement test score, it is the dependent variable on the y-axis, and the aptitude test score is the independent variable on the x-axis. Clerical work should have included recording these scores in the computer files for each person.

1. Correlation is high; we can approximate a straight line in Figure 4–16. Correlation is positive; the line slopes upward from left to right; as one score increases, the other increases; $r = .82$ supports this high positive correlation.

Aptitude Test	Achievement Test
430	200
480	220
480	250
500	200
530	210
530	240
530	240
530	270
550	280
550	290
550	300
560	250
570	270
610	300
640	280
670	270
670	320
720	370

Table 4-3 Scores on Tests

2. r squared = .67 means 67 percent of the variation in achievement tests, y, can be attributed to the aptitude tests, x; 33 percent is caused by other factors.
3. b = .5 means y will increase one-half unit for each unit change in x (in Table 4-3, most of the achievement test scores are approximately one-half the aptitude test scores). We have a = −3 as the value of y when x is 0; in this case, a has no interpretative value since we have no negative scores, and we are not interested in any scores of 0. The regression equation predicts future y scores if x is 500, for example:

$$y = a + bx$$
$$y = -3 + .5(500)$$
$$y = -3 + 250$$
$$y = 247.$$

Everyone who scores 500 on the aptitude test will not score exactly 247 on the achievement test. This is an estimate of future

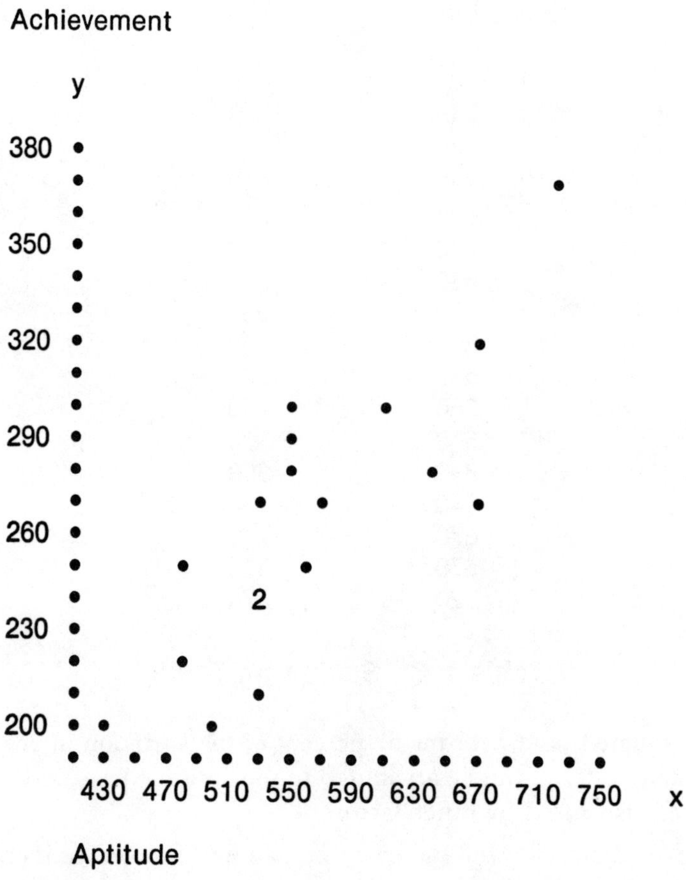

Correlation, r = .82 r squared = .67
Intercept, a = −3 Slope, b = .5
Standard error of estimate = 26

All results are rounded. The computer can produce many decimals but, for most studies, one or two is enough accuracy.

Figure 4-16 Scatter Diagram of Scores

data based on past data, and normally we are interested only in a range, particularly a pass or fail (acceptable or not acceptable) range. We want to know if certain people have a good chance of succeeding, or if they will most probably fail.

4. The standard error of estimate was 26. If data is approximately normally distributed, this means over half, approximately 68 percent, of the future students scoring 500 on the aptitude test should score, on the achievement test, between 247 − 26 and 247 + 26, or between 221 and 273. If less precision is acceptable, we can estimate that 95 percent of the future students scoring 500 on aptitude should score between 247 − 2(26) and 247 + 2(26), or between 195 and 299 in achievement. The modified number line is used again to show the concept of standard error of estimate.

```
--I----I----I----I----I----I----I--
 169  195  221  247  273  299 325
 -3   -2   -1    0    1    2   3
           I---68%---I
      I-------95%-------I
 I------------99%------------I
```

S.E.E. is similar to standard deviation in section 3.3, and cannot be relied upon if the data is highly skewed. If S.E.E. had been 0 instead of 26, then we could predict the achievement test score exactly (an unrealistic situation). Of course, predictions have to be within the limits of the data; we cannot predict an impossible score.

All the statistics we have covered in this chapter cannot replace common sense and experience, but together, they can be valuable in planning and decision making.

5

Showing Differences

1. DETERMINING STATISTICAL DIFFERENCES

Chapter 4 was dedicated to showing *relationships* by comparison. In this chapter we will cover some methods of showing *differences* by comparison. In section 3.3, we talked about differences, or variability, in sets of data based upon averages and standard deviation. We know two groups of people are about equal on a comparable measurement if the corresponding means and standard deviations of the two groups are close. Otherwise, they are considered different. If we are considering only one group, we know the people in the group are about equal on the measurement if the standard deviation is small (most of the scores are clustered around the mean). If the standard deviation is large, we know there is much variability, or inconsistency, in the scores or measurements (many of the scores were far below or far above mean average). Again, the decision about whether a standard deviation is small or large is a comparison based upon past data of the same type, and the size and range of the original values. Much of statistics is based on determining whether characteristics or factors are related to each other. In either case, we also want to know if the relationship or difference happened by chance; was it a coincidence; what are the odds?

In studying differences between factors we have two options—there *is* a difference or there *is not* a difference, a true-false option. The usual method is to approach the problem with the assumption that there *is not* a difference and then try to prove ourselves right or wrong. This may sound like a contradiction, as we would not be considering the situation unless we thought there was a difference; and sometimes the difference is obvious, as in sales this year and last. Our aim is to show a significant difference statistically. We will have this **significant difference** if sets of comparable data differ from each other more than might be expected by chance or coincidence.

The use of computers enables us to have easy access to more data than was possible in the past. In many cases we have all the data necessary to evaluate and make a decision. However, even with computers, it is not always possible to obtain more than a small part of the data we need. We cannot test a drug on the whole population, nor crash all new cars to test safety technology, nor survey the whole population for a public opinion poll. We have to base our claims or decisions on a sample set of data. Sampling theory is a large area in the world of statistics, but it is unrelated to the major purpose of this book.

Before we tackle the statistics of this chapter, let's list some examples of the types of problems in which we would want to show differences—and possible superiority or inferiority: recovery success or effects of drug A and drug B; difference between brand A and brand B; difference between one group with treatment against another without treatment, or one group with special training against another without special training; differences in sales methods or marketing techniques; differences in conditions or attitudes by sex or race, or other factors; before-and-after studies.

There are two sets of conditions under which we test, referred to as (1) groups or (2) pairs. The *groups condition* tests differences between groups as the name implies (income for college graduates against nongraduates). The *pairs condition* tests differences in data obtained from the same person (before and after, pretest and posttest) or the same object, or persons who are alike with respect to certain factors (same education or same IQ, etc.). A familiar example of pairs testing is the study of twins.

2. SHOWING DIFFERENCES BETWEEN TWO ITEMS

One process of showing significant differences employs two sample sets of data and compares the means of those two sample sets. This is called the **t-test**. A limitation is that we have to have two sets of continuous data that are approximately normally distributed. This is not a serious drawback, as much data is continuous, albeit over a certain limited range: income, age, years of education, etc. (section 1.3).

Some of the statistics produced by the computer for the groups t-test include the mean of each group, the mean difference (obtained by subtracting the mean of one group from the mean of the other group), the standard deviation, and a t value. We covered the mean and standard deviation in Chapter 3. The t value is based upon the mean difference and standard deviation (variability) in each group and the number of cases in the groups. The numerator of the formula for calculating the t statistic is the mean difference between the two groups. If there is no difference in the groups we would expect their means to be equal or very close. If the means of the two groups are equal, their difference will be 0; thus, the t value will be 0. The t value becoming larger indicates the possibility that the two means differ from each other more than might be expected by chance, i.e., there might be a significant difference between the two groups.

The F statistic, which will be covered in section 5.3, is used to tell us if this is an appropriate use of the groups t-test. F must be less than or close to 1. Then we know that the two groups have standard deviations that are very close; we are not comparing groups so extreme that the differences shown would be misleading.

Another number, the degrees of freedom (DF), is produced by the computer mainly for information or statistical research reporting. It is based upon the number of cases and is used to determine the **probability**—how confident we can be that the means of the two groups are different enough to conclude that there is a significant difference between the two groups. Statisticians have shown that we get a better estimate from a sample set of data if the degrees of freedom number is used in calculations rather than the actual number of cases. For the t-test, degrees of freedom will be one less than the number of cases ($n - 1$); the $n - 1$ is necessary because of the need to have more than one number to calculate a standard deviation. If we have only one number, the mean is that number; there is no variability; the standard deviation is 0. Degrees of freedom implies freedom to vary; 1 is subtracted from the number of cases, as a set of data having only one number cannot vary.

There is bound to be some error when we are using data only from a sample group. The measurement of error plus the measurement of confidence (or certainty) equals 100 percent. This margin of error is also referred to as the probability or the level of significance. Briefly, if we want to be 95 percent certain that we have a good estimate, the probability should be .05; that is, the level of confidence is 95 percent and the level of significance or probability is 5 percent; if we want to be 90 percent certain that we have a good estimate, the probability should be .10; that is, the level of confidence is 90 percent and the level of significance or probability is 10 percent. One or the other of these will be used in section 5.2 and section 5.3. In section 5.4 we will consider in more detail the limits we will accept when estimating universal results from a small sample.

In considering whether to send our new employees to a specialized school we decided to try an experiment first. Two groups of new employees were given the same test; subsequently one group received on-the-job training while the other group was sent to a specialized school. Groups were then retested with the same test given to both groups. We want to discover if there is a significant difference between learning on the job and specialized schooling. The computer produced Table 5-1 from the test scores. We will not list all the scores as in past examples. The groups t-test does not require that we have the same number in each group. Group 1 has 11 and group 2 has 13 people. Since we have two groups, degrees of freedom is calculated as n − 2 (number of cases minus 1 from each group, or 11 + 13 − 2).

	Number in Group	DF*	Mean	Standard Deviation	F	t	2-tail Probability
Pretest Group 1*	11	22	73.5	9.7	1.38	−.07	.943
Group 2*	13		73.8	11.4			
Posttest Group 1*	11	22	77	10	1.44	−1.89	.072
Group 2*	13		84	8.4			

*Degrees of freedom
Group 1 is on-the-job training
Group 2 is specialized schooling

Table 5-1 T-test Groups

The sign of the t value, + or −, is not important. If the mean of group 1 is greater than the mean of group 2, the t value will be positive. If the mean of group 1 is less than the mean of group 2, the t value will be negative, as in Table 5–1.

An inspection of the means for pretest shows them to be very close, 73.5 and 73.8. The t value, obtained using the mean difference of −.3 as a part of the statistical formula, is low (−.07), very close to 0, or no difference in the groups. For the posttest, the mean difference, obtained by subtracting the means (77 − 84), is larger and the t value is considered larger (−1.89). It is possible that there is a significant difference in the two groups after the training programs.

The standard deviation tells us that most of the scores for group 1 posttest, for example, were between 77 + 10 and 77 − 10, 67 and 87, and certainly between 77 + 2(10) and 77 − 2(10) (section 3.3). There is more variability in the scores for group 1 posttest than for group 2 posttest in which the standard deviation is 8.4. The opposite is true for pretest; there is more variability in group 2 when comparing standard deviations; group 2 is not as consistent. This might indicate that the amount of learning from specialized schooling was more uniform throughout the group (S.D. is 8.4), while on-the-job learning resulted in more unequal training (S.D. is 10). If all those in the group on the job were not trained by the same person, this might be an influence.

In this example, we would like to determine if specialized schooling or on-the-job training was superior. It is evident from the posttest that the mean of group 2 (specialized) is higher than the mean of group 1 (on the job). Chances are the results would be different if we had used *all* our new employees over an extended period of time in this test situation. For now, to be on the safe side, we decide we want to be 95 percent sure we are right in saying that training methods would still show a difference when we used more than just a sample. To be 95 percent sure, we must check the posttest last column, probability, in Table 5–1 and determine if it is less than .05 (5 percent). It is not; it is .072. Therefore, we must accept the possibility that there is not enough difference in the methods to change. Another consideration, of course, is the cost of each method.

The pretest was given to ascertain that there was no significant difference in the groups at the beginning. In Table 5–1, the combination of the low mean difference (73.5 − 73.8), the low t value

(− .07), and the high probability (.943) for the pretest is an indication of no significant difference. A small mean difference indicates that the two sets of data are alike rather than different, that any differences after the training were probably due to the training method. The F statistic close to 1 for both tests shows we have groups that are comparable in both tests.

Two-tail probability, in the last column of Table 5–1, is a term meaning that the margin of error is plus or minus (a possibility of error on either end) as in saying that the selling price is $20,000, give or take a thousand. We have no prior knowledge of which group should be better or worse than the other after training, thus we use what is called a two-tail probability. This is covered in more detail in section 5.4.

Table 5–1 was an example of the groups condition—the analysis of two different groups of people, one group exposed to one condition and the other group exposed to another condition or set of circumstances.

Now let's examine the pairs condition. The pairs t-test implies pairs of scores such as we had in Chapter 4 on correlation. These pairs of scores must be from the same person or object, or from a pair having some close connection. Some of the statistics produced in the groups t-test are also a part of the pairs t-test—DF, the mean, standard deviation, mean difference, t value, and two-tail probability. Correlation, r, is also produced to verify that we have a pair with a close connection. For example, if we compare two groups of higher IQ students, both means are expected to be high; if we compare two groups of lower IQ students, both means are expected to be lower. At least the groups would be matched with respect to the factor, IQ, so that we could test the groups without having the initial IQ influence the results. On the other hand, if we compare a group of higher IQ students against a group of lower IQ students, we do not have a matched pair; we expect a significant difference, one mean is expected to be higher than the other. Thus, the last two groups are not likely to be positively correlated (related). If r is positive, the mean of one group is relatively high and the mean of the other group is relatively high, or both are relatively low. A positive correlation is what we should have to ensure that we are not comparing extreme groups, and will not get a false or misleading significant difference. This suggests that a group of experienced people should not be compared against a group of inexperienced people, if experience might be an initial factor.

We will assume we are considering a work incentive program at extra expense to the company but with the possibility of increased productivity. We choose eighteen employees and test this program for three months. Their productivity per week is recorded and compared against their productivity for the previous three months during which time they were not on the incentive program, a before-and-after analysis. Table 5–2 shows the statistics produced by the computer. Degrees of freedom (DF) is now n – 1, as we have only one group.

The number of mean units produced is larger when employees are on the incentive program, 296.7 compared to 264.4. The standard deviation (variability around the mean) is relatively close for both situations, before and after. The mean difference is simply the result of subtraction of the two means.

R is the correlation we covered previously. For the pairs t-test, correlation between the two sets of data must be positive. If it is not positive, then use the groups t-test because we do not have a matched pair. In this example, we are comparing the same group with itself; there should be *some* relationship if we are making reasonable comparisons. However, the correlation r = .37 in Table 5–2 shows only a small positive correlation. This suggests that the increase in productivity was not uniform throughout the group; the pairs of numbers for productivity before (x) and productivity after (y) would not form a clear straight line on a scatter diagram. It is possible that the more productive employees before the program were already working near their limit and did not improve as much as the less productive employees, or maybe the less productive employees did not respond to an incentive program. To determine this, we would have to look at the actual pairs of numbers for each employee. This could be done by an ordered paired list (such as in Table 4–3) produced by the computer, or by a scatter diagram (such as in Figure 4–16). Age and experience also might be a factor that should have been considered before selecting the experimental group.

The t value in Table 5–2 is higher than the previous example on training methods and the important last column shows we can be almost 99 percent sure that there is a significant difference in productivity—the probability or margin of error is just over 1 per-

	No. of Cases	DF	Mean	Standard Deviation	Mean Difference	r	t	2-tail Probability
On Program	18	17	296.7	46.8	32.3	.37	2.7	.016
Not on Program			264.4	44				

Table 5-2 T-test Pairs

cent (.016). We can feel confident that the work incentive program increased productivity significantly, possibly enough to be worth the extra cost. This now becomes a management decision.

3. SHOWING DIFFERENCES AMONG SEVERAL GROUPS

The widespread use of public opinion polls, preference polls, surveys, evaluations, and questionnaires makes the next statistical method a widely used tool for analysis.

A **contingency table**, sometimes called a **cross-tabulation table**, is the combining of two or more individual frequency tables into a frequency array to compare the frequency distribution of the items in one table with that of the items in the other tables; that is, the counts in frequency tables are shown by groups side by side in one table, or the counts of a single table are broken out into several side-by-side tables for different categories. This tells us that the differences detected are based on *counts* falling into different categories rather than on *measurements* and means as was the t-test. If we want to use these tables with measurement (continuous) data, we can group it into labeled data, for example, income: (1) low, (2) medium, (3) high, instead of using actual income. Categories used frequently with these tables are income levels, age groups, sex, race, education levels, regions.

Suppose, among forty employees, we want to determine if there is a significant difference in attitudes for males and for females. We set up a frequency array (cross-tabulation table) such as that shown in Table 5–3a. The number of males *for* is 8 and the number of males *against* is 12. If there were no difference in attitudes, we would expect the number of females to be about the same as the number of males for each response. We will complete Table 5–3b with the same numbers, 8 and 12 for females, including the totals on the outside and a grand total of 40 employees. The numbers 8 and 12 for females are called expected frequencies; that is, attitudes are expected to be in the same ratio for males and females.

Chi-square is the main statistic computed for these cross-tabulation tables. It is a comparison of actual values with expected values. Since we are looking for differences, it is based on a method of subtraction of actual and expected frequencies, squaring to avoid negatives, and then division. Those who are more mathematically inclined may find the details of these calculations of chi-square in other books on statistics. In Table 5–3b for females, we have filled in the expected frequencies so that the actual and expected frequencies are the same; subtracting 8 from 8 and 12 from 12 will both result in 0, or no difference. Thus, the value of chi-square will be 0, or no difference. We can say attitude is not dependent upon (associated with or related to) sex.

Statisticians have set a frequency of less than 5 as a point in which the calculation of the chi-square statistic would give an inflated value. Either by chance or by educated guesses, some data is expected to be in every category when we are comparing the data (the conditions or the characteristics), especially if we are comparing only two conditions in a 2-by-2 table. If not, then we might not have enough data or we might not have a representative sample (if we have too many of one group and not enough of the other group) to make a valid claim. For example, if we are comparing brand A to brand B, and no one in our survey bought brand B or brand B was unavailable in certain sections, then we are without the data needed to compare the two brands. To make a comparison we *expect* to find some people buying brand B. Therefore, no group should have an expected frequency of less than 5. This means that, for a 2-by-2 table such as those in Tables 5–3a and 5–3b, we should have at least 20 cases if the counts are expected to be evenly distributed over the table. If we have expected frequencies of less than 5, we can regroup the category so that the minimum is satisfied, such as income: (1) less than $30,000, (2)

	Males	Females
For	8	
Against	12	
Total	20	

Table 5-3a Cross-tabulation—Actual Frequencies

	Males	Females	Total
For	8	8	16
Against	12	12	24
Total	20	20	40

Table 5-3b Cross-tabulation—Expected Frequencies

$30,000 or more, or we can get a larger sample of data. As in the past, we use common sense about the way the data is grouped.

Suppose the table appeared as in Table 5-4, this time considering seniority instead of sex. As before, we expect the frequencies to be about the same, proportionately, if there is no difference in attitude. Those with no seniority do not have the expected frequencies when compared to those with seniority; the actual frequencies for no seniority are 5 and 15 instead of 12 and 8 in Table 5-4. Subtracting each expected frequency from each actual frequency (12 – 5, etc.) would result in differences and produce a chi-square greater than 0, and possibly very large. The greater the difference between the expected and actual frequencies, the larger chi-square becomes.

The preceding discussion was a simple illustration of the concept of determining differences based on actual and expected counts for a 2-by-2 cross-tabulation table. To simplify the illustration, the expected count in the second column was based on the count in the first column. The actual and expected counts in the rows are also considered. We will leave both the row and column calculations to the computer, not using the method in the illustration, but using a statistical formula that considers the data as a whole.

	Seniority	No Seniority	Total
For	12	5	17
Against	8	15	23
Total	20	20	40

Table 5-4 Cross-tabulation Illustration

The degrees of freedom figure is dependent on the size of the table, unlike the examples in section 5.2 on t-test; those figures were dependent on the number of cases. The concept of degrees of freedom for cross-tabulation tables can be understood by looking at another 2-by-2 table. The concept is the same for those tables with more than two rows and two columns. Assume we know the total amount of data we have, along with the row and column totals. (The row totals represent one simple frequency table and column totals represent another simple frequency table.) Once a square is filled in with a number, such as the number 12 in Table 5–5, the rest of the squares are determined by the total amount of data. The other three figures will be fixed. Thus, we have one degree of freedom. So far, we have seen only 2-by-2 tables, two rows and two columns. For any number of rows and columns in a table, the degrees of freedom can be determined by number of rows minus 1 multiplied by number of columns minus 1, or $(r - 1)(c - 1)$.

Table 5–6 shows a simple example in which the relationships can be seen easily by looking at the counts in each square (cell) of the table. The totals for each row and column are outside the large square. It shows 18 people received drug A and 15 improved while 3 did not; 12 people received drug B and 2 improved while 10 did not. Drug A is associated with improvement and drug B is associated with no improvement. We might claim that there is a significant difference between the drugs, but not until we look at the statistics produced. A large value for chi-square and a small significance (or probability) show that a difference exists. Significance of .001 in Table 5–6 tells us that there is only 1 chance in a 1,000 of our being wrong in claiming the differences between the drugs. DF and the calculated chi-square are used in determining the significance (see section 5.4). If the computer produces an adjusted chi-square for a 2-by-2 table, this figure should be used, but the significance is the important figure.

	C	D	Total
A	12		17
B			23
Total	20	20	40

Table 5-5 Cross-tabulation—Degrees of Freedom

	Drug A	Drug B	
Improved	15	2	17
Not Improved	3	10	13
Total	18	12	30

DF = 1
Chi-square = 10.46
Significance = .001

Table 5-6 2-by-2 Cross-tabulation

In the next example, we wanted to prove that salary was independent of sex—no significant difference in salary for males and females. Salaries were separated into categories according to what we believed to be low, medium, and high salaries. The data was separated automatically by the computer into these categories after we specified the upper and lower limits for each. In Table 5-7 the top figure in each cell is the count of the number of people falling into that group, and the bottom figure is the percentage (reading horizontally) falling into that group. The column totals show 22 males and 18 females

	Sex		
Salary	Male	Female	Total
Low	4 33.3	8 66.7	12 30.0
Medium	12 66.7	6 33.3	18 45.0
High	6 60.0	4 40.0	10 25.0
Total	22 55.0	18 45.0	40 100.0

DF = 2
Chi-square = 3.37
Significance = .19

Table 5-7 3-by-2 Cross-tabulation

and a total of 40 employees; 55 percent were males and 45 percent were females. The row totals specify the number and percentage of employees in each salary level. For example, for the high-income group, 60 percent were males and 40 percent were females and 25 percent of the whole group were in the high-income bracket. Chi-square is relatively low, denoting a lack of relationship between salary and sex, and any statement of 95 percent certainty cannot be made; significance is not less than .05. Therefore, salary is independent of (not related to) sex for this set of data. We can say there is not a significant difference in salary between males and females.

Another example using the same data but showing a larger table bases the comparison on job type. In Table 5–8, we see there are no supervisors in low- or medium-salary levels and no custodians at high-level salaries. Already we can see there is some relationship between salary and job type. Chi-square of 27.17 is large, denoting a relationship between the two factors, salary and job type; and level of significance is extremely low, meaning we can be very confident in claiming a relationship between salary and job type. We can say there is a significant difference in salary based on job type. The figure of .000 for significance, produced by the computer, could be .0001 or less. This means that there is only 1 chance in 10,000 of our being wrong in claiming the difference in salary based on job type (or the relationship between salary and job type).

As the data is the same as that in Table 5–7, the total number of employees in each income level is the same. We have expected frequencies (the top number) of less than five for each cell, but some statisticians claim this is permissible for tables larger than 2-by-2, if most of the cells do not have fewer than five. We also may claim that the supervision and custodial categories are what we expected. If the counts were expected to be evenly distributed, we should have at least five for most cells.

For Table 5–8, the column percentages have been included as the third number in each cell, and the total percentages as the fourth number in each cell. These cross-tabulation tables often are used only for finding percentages. We are not always interested in determining significant differences. In this way, also, it is similar in use to the simple frequency table (section 2.1). Instructions for reading any cross-tabulation table will be the same as for Table 5–8:

Salary	Job Type Super-vision	Produc-tion	Cler-ical	Custo-dial	Total
Low	0 0 0 0	2 16.7 11.8 5.0	6 50.0 46.2 15.0	4 33.3 80.0 10.0	12 30.0
Medium	0 0 0 0	11 61.1 64.7 27.5	6 33.3 46.2 15.0	1 5.6 20.0 2.5	18 45.0
High	5 50.0 100.0 12.5	4 40.0 23.5 10.0	1 10.0 7.7 2.5	0 0 0 0	10 25.0
Total	5 12.5	17 42.5	13 32.5	5 12.5	40 100.0

DF = 6
Chi-square = 27.17
Significance = .000
If the percentage totals are not exactly 100%, the discrepancy is because of computer rounding.

Table 5-8 3-by-4 Cross-tabulation

1. The first numbers in each cell, adding horizontally or vertically, equal the total outside the cells (actual counts of people).
2. The second numbers in each cell, adding horizontally, equal 100 percent.
3. The third numbers in each cell, adding vertically, equal 100 percent. (See number 6 following).
4. The fourth numbers in each cell, adding horizontally or vertically, equal the total per cent outside the cells; all of these fourth numbers together equal 100 percent.

5. The outside totals percentages, adding horizontally or vertically, equal 100 percent.
6. If the percentage totals are not exactly 100 percent, the discrepancy is due to rounding. (Section 2.1.)
7. Reading one cell of the table, for example, in low-salary custodial there are 4 employees, and the percentages indicate that 33.3 percent of the low-salary employees are custodians, 80 percent of the custodians are low-salary employees, and 10 percent of the whole group are in the low-salary custodial group.

The last major statistic considered in this book is an extension of the t-test to more than two sets of measurements. The **analysis of variance,** or **anova,** can be used for two or more groups, but is usually recommended for three or more groups, with the t-test recommended for only two groups, or two characteristics within the same group. The principles are similar, but instead of a t statistic, the computer will produce an F statistic which shows, in effect, if sets of data are alike or different. As in the t-test, we must have continuous data in which the means of the sets of data can be compared. For example, we could add a rating of job success to the types of training shown in Table 5-1. Then, we could determine if job success is dependent on the training program—on-the-job training or specialized schooling. Using other employee information, we could compare job success by background of employee—business, or science, or arts. For the drug example in Table 5-6, we could add length of treatment in days or weeks for each person, and have a continuous variable with which to determine if length of treatment has an effect on the results.

Anova can be explained more clearly with the use of a Table, 5-9 (not computer-produced), showing the principles of comparing data by analysis of variance. If numbers were in the cells they would be based on means rather than on counts as in cross-tabulation. However, they are not necessary for the initial explanation. The important concept is the way sets of data are regrouped by the computer. Then the data is compared by rows (race), by columns (sex), and finally as a whole group. Of course, anova will compare more than two races, but let's keep it simple for the illustration.

Next, we will put some numbers in the cells for Table 5-10. They represent means for the different groups with each group mean on the outside. G is the grand mean for the groups as a whole. H is the

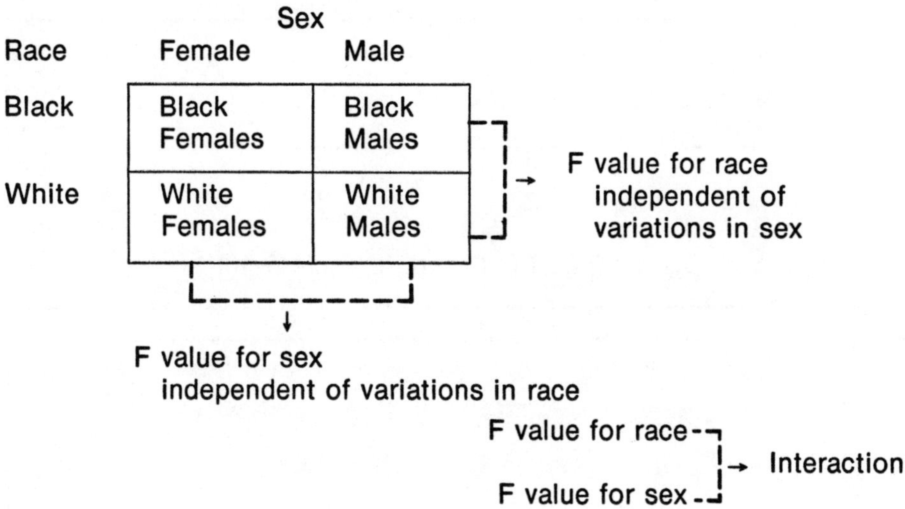

Table 5-9 F Statistic

Race	Sex: Female	Sex: Male	
Black	H = 20	I = 11	N = 17
White	J = 18	K = 13	O = 15
	L = 19	M = 12	G = 16

Table 5-10 Analysis of Variance Illustration

mean for all black females, I is the mean for all black males, etc. L is the mean for all females disregarding race. M is the mean for all males disregarding race. N is the mean for all blacks disregarding sex. O is the mean for all whites disregarding sex. Anova compares L and M to see if there is a significant difference in sex; it appears there might be. Anova compares N and O to see if there is a signifi-

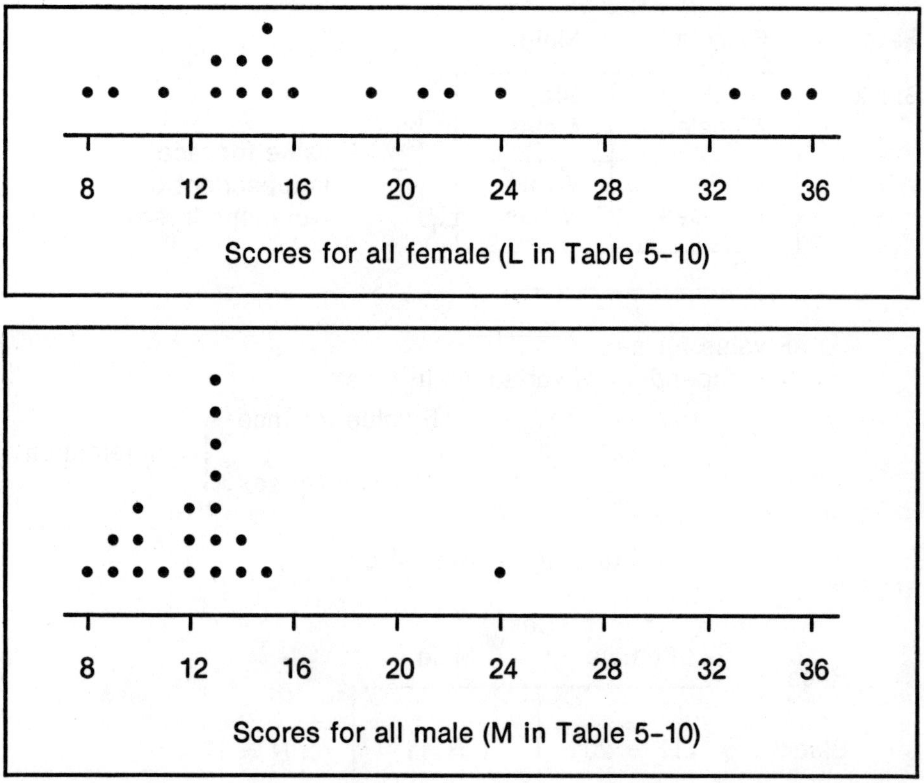

All female:	Mean = 19	S.D. = 8.5
All male:	Mean = 12	S.D. = 3.2

Figure 5-1 Illustration of Variability for Sex

cant difference in race; it appears not. Figure 5-1 shows this comparison for sex (L and M) and Figure 5-2 shows the comparison for race (N and O). Each dot represents one person who made the score listed (or implied) below it. In Figure 5-1, there is a relatively large difference in the two means, all female versus all male, and more variability in the scores shown for females than for males. The scores for males are more clustered together and have a smaller standard deviation, denoting less variability. In Figure 5-2, there is less difference between means and between standard deviations, and although there is a larger number of cases for the all-white group, both sets of scores

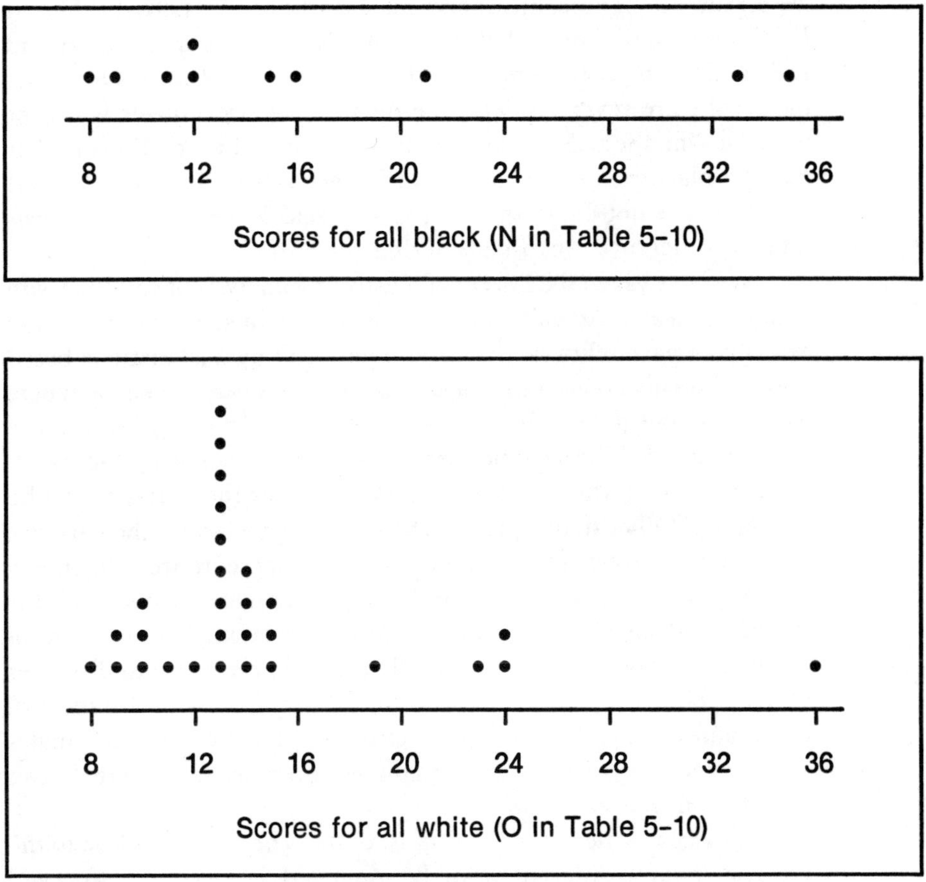

All black:	Mean = 17	S.D. = 9.1
All white:	Mean = 16	S.D. = 6.8

Figure 5-2 Illustration of Variability for Race

are spread out between the score of 8 and the score of 36; one set is not much more variable than the other.

For interaction between race and sex in Table 5-10, H and I are compared, J and K are compared, H and J are compared, I and K are compared. If the differences between H and I, and the differences between J and K, or between H and J, and I and K are about the same, we have no significant difference for interaction; otherwise, we do have a significant difference. The difference between H and I is 9, and the difference between J and K is 5, or 4 more for H and I. The difference between H and J, and I and K is 2 each. However, the results are reversed here; the mean for black females is higher than for white females, whereas the mean for white males is higher than for black males. The largest difference is between L and M, but interaction is not considered here, only sex. We might have a significant difference when considering sex.

We have taken the liberty of illustrating analysis of variance with a table of means. Actually, anova considers means, sample sizes, and amount of variability in the data. The resulting F statistic is based upon variability about the means, thus the term *variance.* The **variance,** like the standard deviation, is based upon the distance of each score from the mean. The variance can be obtained by squaring the standard deviation. Either the standard deviation or the variance can be calculated. Other than squaring the standard deviation, the variance can be obtained by subtracting each individual score from the mean of the group, both within the individual group and between all the groups, and then squaring and adding. The result is called sum of squares. We encountered this in Chapter 4 in reference to distances of data points from the regression line. Now we look at distances of data values from the mean. In Figure 5-1, the data for all females is more variable because there are more scores farther from the mean than for all males.

If there is little variability, most of the scores will be close to the mean so that the differences will be small and the sum of squares will be small. The group will be consistent. Extreme scores far away from the mean, or data points that are more spread out, would produce a larger difference when subtracted and squared, resulting in a larger sum of squares. There would be more variability, or a larger variance. If the groups are not the same size, the larger group will have more scores to subtract and square the difference. It follows that it will have a larger variance when these additional squared differences are added.

Table 5-10 was shown with mean averages in the cells. The final calculation of the F statistic involves a comparison of the variance (variability) *within* each group against the variance *between* all the groups. Considering Table 5-10 again, we can think of the variance *within* each group as the variance of all the scores that fell within group

H, then the variance of all the scores that fell within group I, etc.; that is, we are looking at each cell as a separate unit. We can think of the variance *between* groups as the variance between L and M, then N and O in Table 5–10, and also the columns and rows in Table 5-9. The F statistic is a ratio in which the variance *between* groups is divided by the variance *within* groups. If these two numbers are equal, the result of the division will be 1. Then, we can say there is no significant difference in the scores of any of the groups. If the variance *between* groups is greater than the variance *within* each group, the F statistic will be greater than 1 after the division. In other words, something is causing a difference when comparing one group with another (e.g., L and M in Table 5–10 and Figure 5–1) that is not present within each individual group (or cell of Table 5–10, for example). F being close to or less than 1 indicates similarity in data or comparable groups (no significant difference; the groups being studied are not different). F statistic much greater than 1 (more variability between groups than within groups) causes us to suspect a difference. How much greater depends upon how much difference.

When you punch the computer buttons to get an analysis of variance on data, you may get a format similar to that shown in Table 5–11. The data could represent any measurement data, such as productivity, absenteeism, scores, sales, expense accounts, heart rate, or other physiological measurements. We will use the familiar groupings of sex and race, and the same data used in Table 5–10, for this first example of analysis of variance. The first two rows for degrees of freedom (DF) in Table 5–11 indicate how many separations are within each group—1 less than the number of separations. For sex we have two separations, male and female, and for race we have two separations, black and white. The third DF for interaction is the product of the first two. (This will not always be obvious in the results as all possible interactions will not be listed). The DF for total is number of cases minus 1. We had data measurements from 39 people. The DF for error is that portion not accounted for in the first three, since the total DF is the sum of those above it.

For some groupings, the total sum of squares will be the sum of those above it. In Table 5–11 the sum of squares for sex (based on L and M in Table 5–10 and Figure 5–1) is much larger than the sum of squares for race (based on N and O in Table 5–10 and Figure 5–2). This is an indication that there is more variability (more scores

Source	DF	Sum of Squares	Mean Square	F	Significance
Sex	1	309.1	309.1	7.61	.01
Race	1	4.0	4.0	.1	.76
Interaction	1	24.3	24.3	.6	.45
Error	35	1421.4	40.6		
Total	38	1806.9	47.6		

Table 5-11 Two-way Analysis of Variance

farther from the mean) for sex than for race. The mean square is the sum of squares divided by DF for each item. Division by 1, of course, produces the same number. Each F statistic is obtained by dividing each mean square by the error mean square; 309.1 divided by 40.6 equals 7.61; 4.0 divided by 40.6 equals .1, etc. The error sum of squares is that which was referred to as variance within (H, I, J, K in Table 5–10). You may see the error referred to also as residual sum of squares.

DF, sum of squares, and mean square are used in the calculations for F and then produced by the computer for our information. The last column, significance, is again the important column. With a significance of .01 for sex, we can be 99 percent confident that there is a significant difference in the data when comparing one sex with the other. F, much greater than 1, also shows this. Comparison by sex was illustrated in Figure 5–1. For the comparison of races, illustrated in Figure 5–2, there is no significant difference. The figure, .76 in Table 5–11, is not less than the .05 we have been using. The same is true for interaction of .45. The last two statistics, .6 and .45, in Table 5–11 show that, although there is a significant difference in data based on sex, race is not a contributing factor; the difference is present regardless of race and the lack of difference in races is true regardless of sex.

Table 5–12 is a similar example, showing significance of all items. The computer has processed one set of sales data in which each record of sales had a code for promotion sales and a code for region. Type of sales (promotion or nonpromotion) is called a factor and region is called a factor. These sales records were separated by the computer into four groups using the principle shown in Tables 5–9 and

Source	DF	Sum of Squares	Mean Square	F	Significance
Type of Sales	1	269.4	269.4	7.96	.01
Region	1	158.3	158.3	4.68	.04
Interaction	1	106.0	106.0	3.13	.09
Error	35	1185.3	33.9		
Total	38	1806.9	47.6		

Table 5-12 Analysis of Variance—Sales by Promotion and Region

5–10 so that we have a 2-by-2 factorial analysis. Promotion sales were compared with nonpromotion sales. Sales by region were compared, then promotion sales were compared with nonpromotion sales by region, i.e., promotion sales in region 1, promotion sales in region 2, nonpromotion sales in region 1, and nonpromotion sales in region 2.

The last column in Table 5–12, significance, shows a significant difference in type of sales by promotion and nonpromotion, and also by region. Both are less than .05. If we are willing to accept a level of confidence of 90 percent, then the interaction is significant—less than .10. This interaction, at the 9 percent level of significance and 91 percent level of confidence, indicates that the effect of a promotion was probably not the same for both regions. To determine which is better, if we don't already know, we can use the t-test. This t-test would separate the set of data into two groups: promotion and nonpromotion, and give us that analysis; then take the same data and reseparate it into region 1 and region 2. If we have more than two separations, we can use the t-test only if the separations are tested two by two, i.e., region 1 and region 2, then region 1 and region 3, etc. There are other statistics (not included in this book) that would give us more information, and could be used with the advice of a statistician.

The degrees of freedom (DF) in Table 5–12 indicates how many separations are within each group—1 less than the number of separations (or levels). For type of sales we had 2, promotion and nonpromotion, making DF = 1. For region, we had 2 regions. If we had 3 regions, then DF would be 2 for region; for 4 regions, DF would be 3. Total DF is 1 less than the number of cases (n − 1), sales records in this example. Therefore, we know the computer found 39 sales records (salespersons or stores, etc.).

The sum of squares, mean square, and F were explained in the last example. This example, in the last two columns of Table 5–12 shows how the level of significance decreases, thus the level of confidence increases as F increases for the same number of degrees of freedom. Other souces that may be in your computer file records could have been included, such as type of contact—personal visit, phone, or mail. For this we would have received a three-way interaction: type of contact, type of sales, and region. (More complicated statistics will not be covered in this book.)

4. SETTING A LIMIT

It is assumed in this chapter that we will never have all the data we need to make an unqualified statement or prediction. The computer produces the results after processing the data with formulas that are a part of the computer statistics package. The formulas consider the data based on how much, how it looks, and how it varies—in particular, frequencies, means, and standard deviations. The appropriate formulas produce statistics and a significance. Levels of significance commonly used are .001, .01, .05, and .10. The significance level often used is .05, but this depends upon how confident you want to be that you are right. A level of significance, or probability, of .05 is equivalent to a level of confidence of .95 (95 percent). A result is statistically significant if it is unlikely that it was due to chance. A level of significance of .05 means the result would happen by chance only 5 out of 100 times (once in 20 times); .01 only once in 100 times; .001 only once in 1,000 times. The computer may produce a significance of .000 which can be interpreted as: the result would happen by chance only once in 10,000 times. This is an extremely high level of confidence in being right about your conclusions. As the level of significance goes down, the level of confidence goes up; we can be more confident in claiming that there is a difference in the groups, or factors, we are comparing.

Figures 5–3 and 5–4 may provide more insight into the influence of DF and the value of the statistic on significance. DF is based on the number of cases and/or number of groupings. The value of the statistic is derived from calculated differences. These two figures

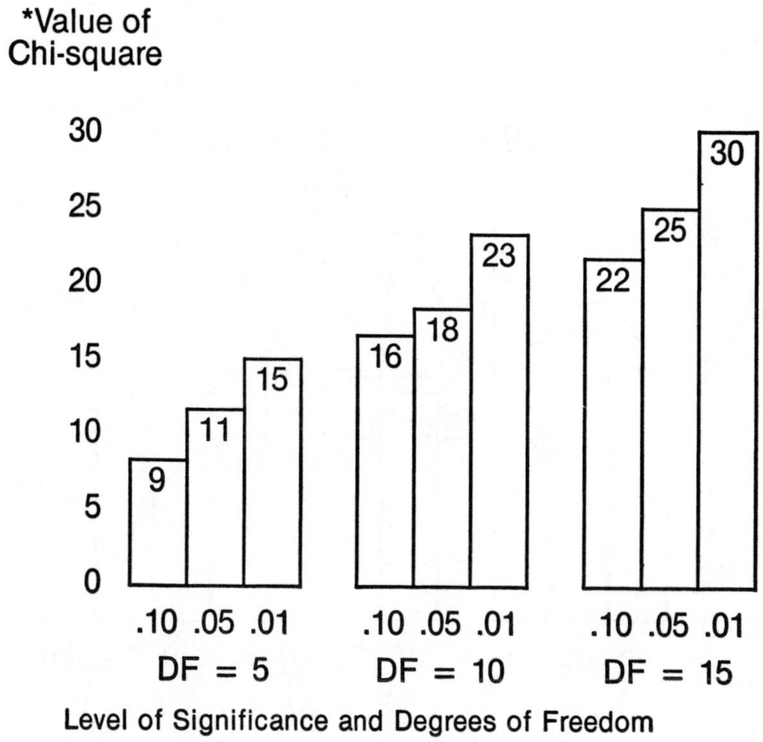

*Does not show decimal portion (rounded)

For a smaller cross-tabulation table (DF = 5) and 90 percent confidence, chi-square must be at least 9, and for 99 percent confidence, chi-square must be at least 15.

For the larger table (DF = 15) and 90 percent confidence, chi-square must be at least 22, and for 99 percent confidence, chi-square must be at least 30.

Figure 5-3 Comparison of Chi-square, DF, and Level of Significance

also show how the value of the statistic must increase (greater difference) for a lower level of significance, thus a higher level of confidence. The degrees of freedom has a different effect upon chi-square in Figure 5–3 than it does on the t value in Figure 5–4. However, chi-square DF was obtained from number of groupings (size of the

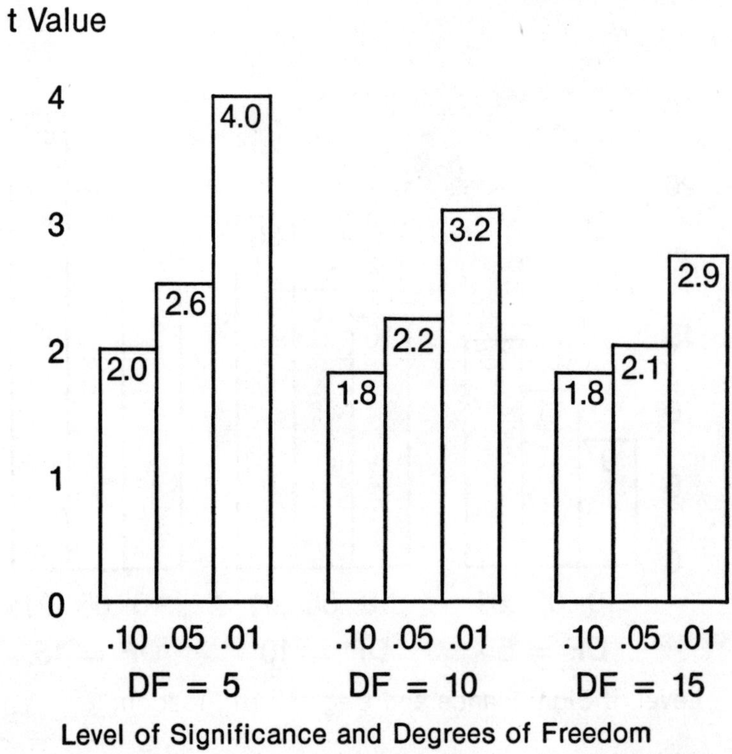

For fewer cases (DF = 5) and 90 percent confidence, t must be at least 2.0, and for 99 percent confidence, t must be at least 4.0.

As the number of cases increases to DF = 15, and 90 percent confidence, t can be as small as 1.8, and for 99 percent confidence, t can be as small as 2.9.

Figure 5-4 Comparison of t Value, DF, and Level of Significance

cross-tabulation table by rows and columns); the DF for t was obtained from number of cases. For chi-square, the smaller the table, the smaller the value for chi-square necessary to be significant. For t, the fewer the cases, the larger the t value has to be in order to be significant. Notice also that the lower the significance, the larger the value of the statistic has to be, as we go from .10 significance down to .01 significance (10 percent down to 1 percent). For illustration, both figures show only three levels of significance—10 percent, 5 percent, and 1 percent, and three degrees of freedom—5, 10, and 15. The computer calculates all of them.

It is left to you to decide on a limit to accept. After deciding on this limit, what you are saying is: if the significance level produced by the computer is greater than this limit, there is no difference in the sets of data; if the significance level is less than this limit, there is a difference in the sets of data.

You have seen the statement—the margin of error is + or - 5 percent, but this is a total margin of error of 10 percent. This can be compared to the two-tail probability in section 5.2. Half of this total, 5 percent, is referred to as a one-tail probability. For example, some jobs have a limit on height. If employers refuse someone who is less than a certain height or more than a certain height, this is a two-tail probability. If there is a limit only on one end, this is a one-tail probability. Normally the two-tail probability is used with the t-test—unless we are interested only in the results on one end, such as proving that brand X is better than brand Y, not just different; or proving that income for college graduates is better than for non-college graduates. Then, instead of a .05 level of significance, for example, we would require a level of significance less than .025 (half of .05), or instead of .01 we would require .005 (half of .01).

The two-tail probability, also referred to as a level of significance, can be shown as the percentage of chance that data is expected to be outside of certain ranges, and the level of confidence is the percentage of chance that data is expected to be within certain ranges, such as

-----.025----- I	------------.95----------- I	-----.025-----
probability of 2½%	level of confidence of 95%	probability of 2½%

The following example is not meant to suggest that teachers should be evaluated on the basis of test scores only; it is simply a description of the use of significance levels. Suppose we are evaluating two teachers by comparing their two groups of students against each other with the use of standardized test scores after the course. Further, suppose that we have decided on a significance level of .10 (10 percent) and the results showed .09. This is less than our limit, so we claim there is a difference between the two teachers; however, we cannot say that one is better than the other at the limit we chose. This now becomes a one-tail probability with a limit of .05 (one-half of .10).

To sum up, we decide (1) on a limit or level of significance (how right or wrong we can afford to be), and (2) if it is one-tail (better or greater than, etc.), or two-tail (a difference but it does not matter in which direction). There are two ways you can go wrong in deciding on a level of significance to accept. For example, a round of golf is played to decide who will be in a tournament. There is a chance (1) of eliminating those who are good but had a poor round, and (2) of including those who are not as good but had a good round.

Considering the difference between teachers again, assume we did set our limit at .10 two-tail, .05 one-tail; received a computer figure of .04; looked at the mean of both groups of students and claimed teacher A was better than teacher B because the students' scores for teacher A showed a higher mean. The two ways of being wrong are: (1) decreasing the level of significance to .01 for a more confident conclusion, and saying teacher A is not better than teacher B when, in fact, he or she might be; and (2) increasing the level of significance and saying teacher A is better than teacher B when he or she might not be. Decreasing the level of significance to avoid one error increases the chances of making the other error.

If the comparison involves money, we have, for example, the problem of accepting or rejecting a group based on the count of defective items in a sample. One error of rejecting too many groups because of a few defective items will lose sales money; the other error of selling too many groups with too many defective items will lose customers. Of course, if all the count or measurement data is included, not just a sample, there will be no chance of making either error.

The decision about what level of significance to choose depends on the type of study, the purpose of the study, how the results will be used, and the cost of duplicating the study—or increasing the sample size—if there is any doubt.

5. SUMMARY

The statistics in this chapter can be used on entire sets of data for the cases in which we have all the data: all our employees or all our students. However, these statistics are used primarily for the cases in which we have only sample sets of data and want to draw conclusions about the whole population or subgroups within the population.

If we receive statistics about one sample, repeat the study on another equivalent sample, then another—the odds are that the results will not be the same. We cannot repeat studies ad infinitum. The level of significance, in effect, figures the odds, or the chances, that different but equivalent sample sets of data will be very similar. We can draw conclusions about whole groups based on the results of samples of the group. The smaller the significance, the safer we can feel in drawing that conclusion.

The particular statistic used will depend upon the kind of data and the number of groups. For two sets of measurement, or continuous, data the t-test with t statistic is used. For more than two sets of this type of data, analysis of variance with F statistic is used. For sets of data that involve counts of how many people or objects fall in each group, the cross-tabulation with chi-square statistic is used.

It is likely that statistics not covered in this book will be a part of some computer statistics packages; it is also likely that the methods for producing the statistics will be different, resulting in minor discrepancies between instruction manuals, but this book is a good beginning.

The more in-depth or more complex studies may require the help of a statistician. At least, with your prior knowledge and some statistical results already obtained, statisticians won't have to start from scratch. They can recommend other statistics that are appropriate to use. They can help with more detailed analysis of the data. They can review the results to ensure that no false conclusions have been drawn because of a mistake in the use of data, sampling, grouping, etc., as it is possible for the inexperienced person to collect data that is virtually worthless.

The analysis of variance, in particular, has been presented only in a broad sense. Understanding the many ramifications of the different versions of anova will require consultation with a statistician. The help we need, hence the price we pay, is minimal when we are ahead on our knowledge of statistics. We're not statisticians; we may be smarter than we were, but we don't know it all. Yet we can feel comfortable using basic statistics and talking with statisticians.

SECTION

II

USING BASIC STATISTICS

6

Developing a Computer Approach

1. USING AVAILABLE COMPUTER RECORDS

The following representation of file folders is meant to convey the idea of computer files. The concept is the same. A computer file is a set of related information. This information may be brief, as in a file of names containing only one entry or name per record (or file folder); or it may be extensive, containing all necessary information about a person or object within the same computer record. Thus one file folder represents one computer *record* of information—all information on *one* customer; and the set of folders (file drawer or file cabinet) represents one computer *file*—all information on *all* customers. Some of the common types of files are shown here.

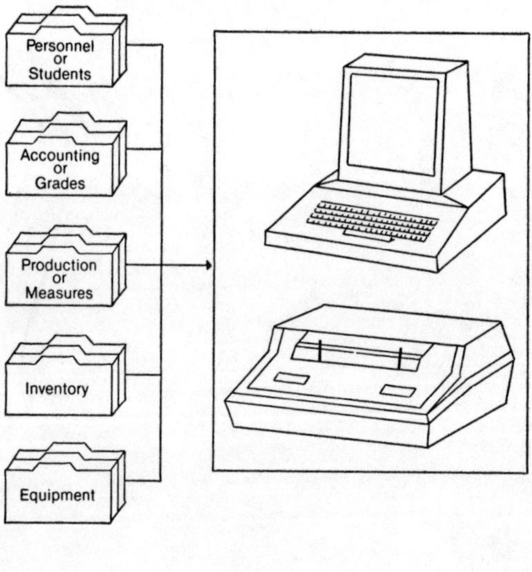

In Table 6–1, we have one file representing equipment (or inventory) with a portion of the information as it would appear in a paper record in a file folder and then two ways it could appear in a computer record. In the column format, the numbers above designate the starting point of each item of information; for example, the serial number is in columns 14–19, and the model is in columns 27–30. Although the model is three characters in this example, other models may be four characters and must be provided for in the column format so that column 30 is used for the four character model. This is also illustrated with the leading zero for purchase order number. In some cases purchase order number can be six characters. In other words, the length of each item depends on the length of the longest possible entry that could be in that item. In the basic format, serial number, 80321, is the second item (variable 2) in the record and model, BG1, is the fourth item (variable 4).

Time wasted is equated with money wasted. High-salaried people may spend time searching through stacks of paper and recording counts of certain information. What a waste of personnel! What a waste of expensive computer equipment! Don't overlook data that is already present in the computer files. The data can be used for purposes other than that for which it was originated. For example, printing payroll checks, with attendant reports, does not have to be the only purpose of salary information. Some people manually compute percentage of pass/fail on standardized tests by counting the number passed and the number failed, person by person. They may know the scores are recorded in the computer but it never occurs to them that it can be used for other than printing on transcripts, that the computer can count the number of passes and the number of fails, person by person, and produce a total number and total percentage along with other appropriate statistics. Know what bits of information are in your computer records. If you don't know, ask the software vendor or someone in your computer department. They cannot be expected to be aware of all your needs for information; you may be the only one close enough to the situation to know what comparisons or statistics would be useful. We have to have communication, or we will have confusion. If data is in the computer for some purpose, a good manager or advisor will know what the data represents and take full advantage of it.

1. Type of Equipment	Refrigerator
2. Serial Number	80321
3. Make	Falcon
4. Model	BG1
5. Purchase Order Number	56900

File Folder Record

```
1            14    20     27   31  36
REFRIGERATOR 80321 FALCON BG1 056900
```

Computer Record Column Format

```
REFRIGERATOR, 80321, FALCON, BG1, 56900
```

Computer Record Basic Format

Table 6-1 Examples of File Records

Related information should be together in a computer record and in a computer file, just as it would be in a file cabinet; that is, all salespersons' records should be in one file and information normally used together should be adjacent in the record. A salesperson's number should be adjacent to codes for state or region or district. They can then be processed together, or easily sorted by state, etc. By codes we mean, for example, "1" for south, "2" for east, "3" for north, "4" for west. The choice of which code you use does not matter but you will want to keep a record of which code, or label, you have assigned to each area. Even though there may be no statistical necessity for order, the frequency table and cross-tabulation table will show the data in some sorted order, either ascending or descending. If you want the north area to be listed first, then it should have the lowest code, if you plan to use the ascending sorted order primarily. Using numeric codes normally shortens processing time and the fewer characters saves valuable storage space in the computer, and facilitates data entry. The more efficient data-entry devices, or keyboards, have all numerics positioned together. It is much faster to enter "1" than to type "south." If you prefer, you can use "S," "N," "E," and "W" since this is only a label for separating data. A computer sort of the records by region normally would produce these in alphabetical order with east first and west last; however, numerical codes are really more efficient.

The computer record in Table 6–1 could be saved with the use of a code for equipment and a code for make, for example, "refrigerator = 042" and "falcon = 103." The record, in column format, then would appear as follows:

```
1   5      11  15   19 24
-------------------------
042 80321 103 BG1 056900
-------------------------
```

Of course, we would have to save the assigned codes and the appropriate names in another file in order to match "042" with "refrigerator" and "103" with "falcon." As you see, if we have many refrigerators in stock we could save much computer space and much data-entry time if we save the word "refrigerator" only once on a record instead of on every equipment inventory record for refrigerators. This also reduces the chances of spelling errors in data entry. For printing reports or lists for which we want the equipment name, we could have the computer pick the name out of the name file by its code number, or by its position in the record (see section 6.3 and Table 6–7).

Buy computers that are capable of combining payroll, accounting, personnel, and all the other functions you need, along with statistics. Buy computer statistics packages that are adaptable to the way the data is recorded in the computer, instead of reorganizing data to fit the computer statistics package. Some statistics packages are designed to be usable with the other software you have purchased. If this is not true, the vendor of your other software should be required to provide you with the formats, or record descriptions, of all files. Then you will know that salary is the fourth item in the payroll file record—or is recorded in spaces 29 through 36—depending on the type of computer. The statistics package should provide an easy way to pick salaries or other data out of the record, by its formatted position, and designate where a decimal point is to be located, if needed. Usually decimal points are not stored in the computer; they are inserted automatically for performing arithmetic or printing. Of course, someone has specified for the computer when defining the data that salary, for example, has two decimal places.

Table 6–2 shows a condensed version of one file, personnel, with some of the items that might be useful for statistics, along with the type of data on the right. They are in order as they would appear in a computer file record. The left-hand column is a simple record description. The first three, of course, are of no use for statistics, although the ZIP code portion of the address may be of interest to companies that fill mail requests. A frequency table by ZIP code will show in what areas most customers are concentrated. Some information we can learn from the remaining items is: (4) how many married or single, (5) how many females or males, (6) how many of each race, (7) how many over or under a certain age, (8) how many with the company more or less than a certain number of years, (8) how many hired, by year of hiring, (8, 9, 10) increase in salary by longevity, (10) how many earning more or less than a certain salary, (11) how many in each department, (12) how many in each job, (14) how many have taken more than a certain number of sick days. You can get averages and percentages of the items.

All this information can also be obtained in groups, such as by department or by sex. You may want to find out which departments have taken the most sick days, or if number of sick days taken is re-

1.	ID Number (unique)		
2.	Name		
3.	Address		
4.	Marital Status	(for statistical use)	Labeled
5.	Sex Code	(for statistical use)	Labeled
6.	Race Code	(for statistical use)	Labeled
7.	Birth Date	(for statistical use)	Continuous
8.	Date Hired		Continuous
9.	Beginning Yearly Salary		Continuous
10.	Present Yearly Salary		Continuous
11.	Department		Labeled
12.	Job Code		Labeled
13.	Sick Days Earned		Continuous
14.	Sick Days Taken		Continuous

Table 6-2 Sample Personnel Record

lated to job. You may want to estimate, from data on ages, how long it will be before you need a significant number of new employees. You may be considering an insurance plan for employees and can afford a certain premium per person depending on age; you can get an estimate of the total cost of the premium. Of course, for statistics on ages, the date of birth of each person will have to be subtracted from present date. The computer can do this easily and "remember" the ages to produce the results you need. This process is facilitated in some computers if birth date is recorded with year and month adjacent instead of using month, day, and year, but it certainly can be done any way.

Grouping data usually is not necessary by computer as a computer can handle massive amounts of different data. If there is a reason for your information needs, it can group age, length of service, salary, or salary increase (for example: low, middle, or high salaries, depending on your determination of the ranges for each). All data should be included within one and only one of the ranges.

All of the information in Table 6-2 can be combined in various ways depending on what you want to see—or what you want to prove. It could be average salary or salary increase compared to length of service, or the same information grouped by job, or by sex, or by race, or by department. You may want to discover if one department is costing you more money than is necessary.

Statistics won't give you all the answers, but they will show if something needs further investigation. Once the reports you need on a continuing basis have been determined, the instructions for obtaining these reports can be saved in the computer and recalled, with very little effort, when you want the reports. This may be once a day or once a year, or every time there are major changes in data. Whenever possible, the information needed should be decided on before using the computer. This saves the time of repeating steps after you have decided you want something added to what you have. *The ultimate aim is to assimilate only the necessary data in the fewest number of reports to provide the information needed to manage or advise effectively,* whether it concerns employees, students, time and money, or objects. The use of appropriate data for effective management often is referred to as management information systems (MIS) or decision support systems (DSS).

The information available in the different types of files, some of which were shown at the beginning of this chapter, will make it possible to produce other statistics. We will mention a few without going into detail—cost/revenue analysis, frequency of deliveries, price increase justification, cost to produce a certain product, number of sales, advertising effectiveness, production quotas, incoming orders by price of product, comparison of product quality or life expectancy, cost of equipment and maintenance time (when to replace equipment), number of customers, work force scheduling, type and number of complaints from customers, turnover in employees by department or supervisor, loss or gain of students by school or department, budget overrun by department, backlog of orders from certain companies (date received minus date placed), defective items received from certain companies, test comparisons, police data, accidents, tickets, agricultural data. Anything in the computer files can be counted or compared, if it is count or measurement data (the majority of computerized data should fit this description). Often *statistics can be thought of as nothing more than rearranged numbers.*

This section has dealt with data that is, or should be, a part of your computer files, if it is applicable to your situation. Additionally, some data-entry machinery can be hooked up to a computer in which measurements are being recorded as they are made. This data, automatically entered into the computer, can be recorded in a manner similar to that of the other files. To obtain statistics from it, we need only know the record description (where each measurement is located within the record or the name of the item). This is mentioned again in section 6.3. Portable computers, or terminals, can receive the data at the source—the field, the checkout desk, the storage bin, the lab, the hospital, etc.

To use computer statistics packages most efficiently, all your computer records should have some kind of code or number for all items in the record; for example, if no parts designated as part number 114 were sold, the record should contain zeros instead of blanks; if no college major has been decided on by a student, that student's record should contain a code for undecided rather than blanks. All files that sometimes have to be tied together to look at certain items of data should have common cross-indexing codes to match. If you have a good computer department, they will take care of this. If not, make sure the software vendor can tell you if and how this can be accomplished. For example, you may want to compare salary with amount of production by employee. The identification number of the employee will be a match on both records, the personnel record and the production record. Then statistics can be produced from information in two or more files about a person, without duplication of information in the several files. A student's grade history file will have an ID number to match one number in each of the other student information files, so that we can have code matching between files.

2. SETTING UP SPECIAL COMPUTER RECORDS

It was assumed in the last section that all the data had been collected, stored in the computer, and updated on an ongoing basis using other computer software. In this section, we will consider the special project, the collection of data for a specific purpose, like research or surveys. The data will be used and discarded, or possibly added to, but not changed. A survey of customers will not change the results of any previous survey. Examples of a survey and a research project are shown in Chapter 7.

Avoid collecting data with the intention of analyzing any data you just happen to get; instead, determine the information you want to analyze and then collect the data. You may regret having omitted the collection of some needed data when it is too late to repeat the same conditions under which you gathered the original data.

After determining the information you want, design a method to capture this information that gives no cause for indecision by the data gatherers or recorders—and a method that allows them to gather and record data quickly. For research and some surveys, forms with clearly marked spaces for recording specific data items should be listed on the form in order of the sequence of gathering so that the recorder will not have to skip around on the form. If data entry is necessary, the person doing this task should be able to enter the data directly from the form, similarly without having to skip around on the form—from the top down as well as left to right. The result is much, sometimes a critical amount of, time saved; and the chance of error in data entry, or in recording data on the forms originally, is reduced.

For other surveys or questionnaire projects, the data should be gathered by multiple-choice questions whenever possible. If data entry is necessary, numeric answers can be typed faster. Questions that request a degree response should be consistent, either from high to low or from low to high; do not use possible choices of (1) always, (2) sometimes, (3) never, in one question, and (1) never, (2) sometimes, (3) always, in another question. This is confusing to the respondent and will make the statistics from the computer more difficult to read. Questions should be worded carefully; they should provide for the choice of a "don't know" answer to distinguish it from the question that was simply left blank, if "don't know" might be an alternative answer to the question. Sometimes the main purpose of a question is to determine how many people don't know certain information. On the other hand, if there is a large percentage of "don't know" answers for questions that should be possible to answer, then the question might be inappropriate or poorly expressed.

If you are fortunate enough to have access to a *mark-sense* reader connected to a computer, it will facilitate these types of studies. This, or a similar data-entry method, is probably the least expensive, least error-prone, and most efficient method of mass data gathering. It bypasses the step of having a person enter the data. The data is read from the mark-sense forms directly into the computer. Examples of one version of the type of forms used in readers are shown in Figures 6–1 and 6–2. In addition to surveys, and questionnaires, we can use forms similar to these for ordering (fill order and reduce inventory at the same time), counting stock, delivery truck records, insurance or other collection records, expense records, field service records, customer contact reporting, utility or police or real estate field recording, or recording any data where it is not practical to have a com-

IDENTIFICATION NO.						SHEET NO. ① ② ③ ④ ⑤ ⑥ ⑦ ⑧ ⑨	
						ANSWERS	
⓪	⓪	⓪	⓪	⓪	⓪	1 ① ② ③ ④ ⑤	11 ① ② ③ ④ ⑤
①	①	①	①	①	①	2 ① ② ③ ④ ⑤	12 ① ② ③ ④ ⑤
②	②	②	②	②	②	3 ① ② ③ ④ ⑤	13 ① ② ③ ④ ⑤
③	③	③	③	③	③	4 ① ② ③ ④ ⑤	14 ① ② ③ ④ ⑤
④	④	④	④	④	④	5 ① ② ③ ④ ⑤	15 ① ② ③ ④ ⑤
⑤	⑤	⑤	⑤	⑤	⑤	6 ① ② ③ ④ ⑤	16 ① ② ③ ④ ⑤
⑥	⑥	⑥	⑥	⑥	⑥	7 ① ② ③ ④ ⑤	17 ① ② ③ ④ ⑤
⑦	⑦	⑦	⑦	⑦	⑦	8 ① ② ③ ④ ⑤	18 ① ② ③ ④ ⑤
⑧	⑧	⑧	⑧	⑧	⑧	9 ① ② ③ ④ ⑤	19 ① ② ③ ④ ⑤
⑨	⑨	⑨	⑨	⑨	⑨	10 ① ② ③ ④ ⑤	20 ① ② ③ ④ ⑤

Figure 6-1 General-Purpose Sample Mark-Sense Form

BIRTH DATE				MARITAL STATUS	○ ○ OFFICE USE
Yr		Mo		○ Single	SEX
				○ Married	Ⓜ Ⓕ
⓪	⓪	⓪	⓪	○ Widowed	WORK BACKGROUND
①	①	①	①	○ Other	○ Health related
②	②	②	②	HIGHEST EDUCATION	○ Business
③	③	③	③	○ Below 8th	○ Trade
④	④	④	④	○ 8th – 12th	○ Industrial
⑤	⑤	⑤	⑤	○ High Sch. Grad.	○ Education
⑥	⑥	⑥	⑥	○ Some College	○ Retail
⑦	⑦	⑦	⑦	○ College Degree	○ Domestic
⑧	⑧	⑧	⑧	○ Graduate Work	○ Other
⑨	⑨	⑨	⑨	○ Graduate Degree	○ Never employed

Figure 6-2 Special-Purpose Sample Mark-Sense Form

puter on site. Even where a computer may be available, it is sometimes easier to use the forms initially. Many schools use them for such things as enrolling students, giving tests and evaluations, and recording grades.

Your forms or questionnaires can then be printed, filled out as needed, and read into the computer with a mark-sense reader. If the cost to purchase the special forms is not justified, an inexpensive general-purpose form, with answer spaces only, can be used. Care must be taken to indicate clearly which spaces on the mark-sense form correspond to the questions on another sheet of paper. Using the usual numbering scheme should present no problem. Examples of both types, showing a portion of a page for each type, can be seen in Figures 6–1 and 6–2 (use a number 2 pencil to fill in the bubbles). Neither shows spaces for alphabetics, such as a person's name, but this is available, if needed.

3. LAYING OUT THE SOLUTION

What do we want to find out? What must we have to find this out? Whether you are using a computer or not, these two questions are the beginnings of an investigation. Let's assume for the first question that we want to determine where to locate a new branch office. Further, assume we have a computer with a statistics package that can provide the count data we want. For the second question, aside from socioeconomic aspects and possible future customers, we decide that one of the influential factors is money—which states have the most significant customer concentration, which states bring us the largest average dollar volume, and possibly which states have the largest freight costs to ship from here to there. Of course, a manager can make a knowledgeable guess about these, even though the summary sales reports may not be broken down by customer.

To simplify the example of obtaining this more accurate information by computer, we will consider only the customer concentration. The manner in which this information is obtained will depend, as always, on the type of computer and statistics package. Usually it is as simple as this:

```
FILE         CUSTOMER
DATA         STATE 30-31
FREQUENCY    STATE
```

These three statements tell the computer that the STATE portion of the customer's address is in columns 30 and 31 of the CUSTOMER file records and we want a count (frequency) of how many customers are in each state. The computer will inspect columns 30 and 31 as one data unit in each record and add each occurrence of data that is alike, as you would do it with pencil and paper—by looking at the state portion of each address in a mailing list and adding each occurrence of AL, AZ, etc. We assume we have valid state abbreviations. (An error in updating customer records, such as a misspelled state abbreviation will be counted as an additional separate state by the computer.) Your computer software for updating customer records should provide a method of checking data to keep bad data from entering the records. The appearance of the results of this frequency by state is shown in Table 6-3. The figure 0.0 for AL indicates that the percentage was less than 0.1. The cumulative frequency column was of no use for this information and was not reproduced in the table, even though the computer produced it automatically as a part of the list.

Obtaining additional information from the customer file, or other data files, is no more difficult, just more time consuming to key in what you want. The first command FILE, can remain the same for whatever customer information is needed and the second command, DATA, can remain the same if all the data in the customer file is defined as the example for state has been, i.e., customer number,

State	Frequency	Relative Frequency Percent
AL	39	0.0
AR	3683	3.0
AZ	296	0.2
•	•	•
•	•	•
•	•	•
WI	6911	6.0
WY	301	0.3
Total	79592	100.0

Table 6-3 Frequency Table—Customers by State

discount rate, amount due, amount paid. Then you can change the third command or add commands according to the statistics you want. A breakdown by region may be wanted. This can be done, the hard way, by defining what states are in what regions, or the easy way, by having a region number as a part of the customer number. For the latter, we would designate where the region number was located in the record and the computer would group all like regions together before producing the report. We could also have grouped the better customers together, using the number of units ordered—or some other measurement of good, average, or poor customers.

For the next example, assume we want to find out how we can improve services to senior citizens. All the information we need to find this out is not in a computer. We decide a survey is the only way. The responses to the survey will have to be entered into the computer by some method of data entry, covered in section 2 of this chapter. For any method, the results will be the same; the data will be in the order that it is put into the computer. In the mark-sense form in Figure 6–2, we have the first five questions of this survey: birth date, marital status, education, sex, and work background. After this and the rest of the responses are in the computer, we can request statistics in the following manner:

```
FILE      SURVEY
DATA      YEAR 1-2, MO 3-4, MARITAL 5, EDUC 6,
          SEX 7, . . . , LAST 58
COMPUTE   AGE = PRESENT YEAR - YEAR
FREQ      AGE, MARITAL THRU LAST
```

The first statement shows that the data is in a file called SURVEY; the second statement, DATA, tells the computer where in the record each response is located. YEAR of birth is in columns 1 and 2, and the others are defined up through the LAST question in column 58. The COMPUTE command is one simple method of calculating age in years by subtracting year of birth from present year, and the FREQ (frequency) gives us a count of responses to all the questions with the percentage of occurrence of each response. Notice that we omitted month of birth (MO) in the request for frequencies. It was requested in the survey so that we have the information if we want it later for a more detailed study. Whether using available computer records or creating special computer records, it is wise to define the location of all the data items in the event that they are needed later. As competent as we are, there are always additional bits of information we find would be useful.

For one of the questions, MARITAL status, column 5 in the computer file records could contain a "0" for single, "1" for married, "2" for widowed, "3" for other, or "9" if blank or unanswered. All other questions would follow the same general pattern for the possible responses. Our determination of some of the needs for services, or for activities, for this group of senior citizens might mean locating various predominant conditions, such as marital status and highest education. For percentages of those in each group, we could use

TABLE EDUC, MARITAL

and receive a cross-tabulation table similar to that in Table 6-4. The column totals at the bottom represent a simple frequency table for marital status, and the row totals on the right represent a simple frequency table for education. Each cell contains a count of the combined vertical-horizontal category. Widowed with 8th–12th education is predominant—364 or 33.1 percent. (See section 5.3 for explanations on interpreting these cross-tabulation tables.)

If, for some reason, we wanted information isolated by marital status, we could use

PICK MARITAL = 0

and receive the statistics only on the single respondents. If we wanted information on all others, excluding the single respondents, we could use

OMIT MARITAL = 0

Various combinations of *picks* and *omits* can be used for each item in the survey. To provide for those unanswered, in the case of marital status, we could use

MISSING MARITAL = 9

In other words, all spaces in the question about marital status were left blank by the respondent. We could exclude those from the statistics or we could receive adjusted statistics including those left blank and coded as a value of 9. If the space (bubble) was filled in, it would be coded automatically by the mark-sense reader as a 0, 1, 2, or 3 and this number would be inserted in column 5 of the computer record for that respondent. Section 7.1 gives a detailed example of the survey type. Other examples of surveys can be found in sections 8.1 and 9.2.

Education	Marital Status Single	 Married	 Widowed	 Other	Total
Below 8th	5 16.7 13.5 0.5	22 73.3 5.2 2.0	2 6.7 0.3 0.2	1 3.3 33.3 0.1	30 2.7
8th-12th	12 2.4 32.4 1.1	121 24.3 28.5 11.0	364 73.2 57.2 33.1	0 0 0 0	497 45.1
High School Graduate	9 4.4 24.3 0.8	109 53.4 25.6 9.9	85 41.7 13.4 7.7	1 0.5 33.3 0.1	204 18.5
Some College	7 4.6 18.9 0.6	62 40.8 14.6 5.6	83 54.6 13.1 7.5	0 0 0 0	152 13.8
College Degree	3 2.1 8.1 0.3	93 64.6 21.9 8.4	47 32.6 7.4 4.3	1 0.7 33.3 0.1	144 13.1
Graduate Work	1 1.4 2.7 0.1	18 24.7 4.2 1.6	54 74.0 8.5 4.9	0 0 0 0	73 6.6
Graduate Degree	0 0 0 0	0 0 0 0	1 100.0 0.2 0.1	0 0 0 0	1 0.1
Total	37 3.4	425 38.6	636 57.8	3 0.3	1101 100.0

Table 6-4 7-by-4 Cross-tabulation

Some computers handle these statistics in a manner similar to that shown in Table 6-5. First, we would select option 4, data management, and receive the menu for handling data. This is shown in Table 6-6. In this menu we will select option 1, define data, and receive the prompts to define the variables.

VARIABLE 1 ?	to which we enter YEAR
COLUMN ?	to which we enter 1 – 2
LEGAL VALUES ?	to which we enter a number range or the word NUMERIC

In some computers we may not need to indicate columns. The result of all the data definition for the previous survey example is shown in Table 6-7, which we will SAVE and use for the data file called SURVEY. Now, we return to the main menu, Table 6-5, and select

Main Menu

1. System Management
2. File Management
3. List Directory
4. Data Management
5. Statistics
6. Utilities
7. Exit (or Quit)

Choice (enter number and hit return) ?

Table 6-5 Example of Main Menu

Data Menu

1. Define Data
2. Create Data
3. Change Data
4. Save Data
5. Examine Data
6. Merge Data
7. List Data
8. Exit (return to main menu)

Choice (enter number and hit return) ?

Table 6-6 Example of Data Management Menu

option 5, statistics. This statistics menu will have a list of the various statistics as shown in Table 6-8. If we wanted Table 6-4, we would choose option 5, cross-tabulation, and enter SURVEY, EDUC, and MARITAL.

Variable	Column	Legal Values
1. Year	1 - 2	numeric
2. Mo	3 - 4	01 to 12
3. Marital	5	0 = Single
		1 = Married
		2 = Widowed
		3 = Other
4. Educ	6	0 = Below 8th
		1 = 8th-12th
		2 = High School Graduate
		3 = Some College
		4 = College Degree
		5 = Graduate Work
		6 = Graduate Degree
5. Sex	7	0 = Male
		1 = Female
•		
•		
•		

Table 6-7 SURVEY Data Definition

Statistics Menu

1. Descriptive Statistics
2. Frequencies
3. Graphs
4. Correlation and Regression
5. Cross-tabulation
6. t-test
7. ANOVA
8. Exit (return to main menu)

Choice (enter number and hit return) ?

Table 6-8 Example of Statistics Menu

For one more example, we want to find out how well our high school is preparing students for college. We need a standard with which to compare. It would be ideal if we could have access to all college records on our past students to compare to college records of other students. Lacking this we have to use those records we have on our computer. This is better than doing nothing. We decide there should be some correlation (Chapter 4) between high school average and college entrance exam scores. Our objective is to find out how good our college prep courses are. Is there a possibility we are giving out unearned grades, or not covering the necessary material, or using less capable teachers for some of the college prep courses? We can obtain the correlation by telling the computer which file and where the data is located in the record and what statistic, in this case, a scatter diagram.

```
FILE     STUDENT
DATA     HSAVG 42-45,CEE 60-62
PLOT     HSAVG, CEE
```

The data is in a file called STUDENT, the current high school average (HSAVG) is in columns 42 through 45, and college entrance exam score (CEE) is in columns 60 through 62. PLOT requests a scatter diagram of these two with the correlation between them. If the college entrance exam was not recorded in a student's file, we assume that student did not take it. The computer would automatically include in the correlation only those students with both measurements—high school average and college entrance exam. We can also compare the mean average of the college entrance exam for our school against any published national or state average, as this statistic for our school should be produced as a part of the plot we obtained.

Another method for comparing internally would be to have the computer separate those students who entered our school as freshmen and those students who transferred into our school. This would compare those who received all of their training in our school with those who did not. This is a groups comparison. The t-test (section 5.2) could be obtained by adding

```
T-TEST   CEE, GROUPS REGULAR AND TRANSFER
```

After adding, or already having, the location of the code for transfer in the DATA statement, the t-test would tell us if there is a significant difference between scores for transfer students and scores for the students who entered our school as freshmen. If our school is part of a larger school system, we can make the comparisons with other schools in the system. We would like to prove what we want to believe—that students spending four years in *our* school are better prepared. If the converse is true, we want to find out why.

The versatility of statistical packages and the ease of use by nontechnical people precludes the necessity of buying or writing a different computer program for all of the information we want. In effect, we write our own list of instructions and save them, if needed, to be used again and again. Although not necessary, you can make your job even easier if you have computer personnel who can prepare all the preliminary statements, in particular the DATA locations, or definitions. Then all you have to do is add the commands for the statistical reports you want.

For that information for which a long list of different names is needed, some statistics packages may be of no use. They are primarily for use with numbers and short descriptive names. The method of obtaining these statistics has been generalized in the examples. The instruction manual with your statistics package should outline simply the directions for obtaining the statistics, and the explanations in this chapter should give you a good start on understanding the instruction manual.

The importance of planning ahead in setting up computer files of information cannot be overstated. The ease of obtaining statistics has been stressed through this book, but the assumption has been made that data is organized efficiently and completely. Many people have the false impression that one button can be pushed and the report comes out. These are the people who have talked to some computer salespeople and probably have little familiarity with a computer otherwise. Someone, somewhere, should have spent many weeks, or months, or even years, ensuring that data will be arranged in a manner that will make it possible for you to let the computer do the work.

Many companies with computer departments have realized the necessity of including computer personnel in upper-level management and planning. The organization and plans of the company or institution have a direct impact upon the organization of the computer department, and of data management methods to satisfy the information needs of all facets of the organization. In addition, computer personnel have to know what might be coming in the future, to be prepared for development and for changes. Otherwise, you will have inefficiency, duplication of effort, and at worst, "end up with a centipede when a biped will do." You will end up working for the computer instead of having the computer working for you. The key message is: you have to be organized *before* you can be computerized.

7

Health and Human Services

1. JAIL SURVEY

This jail survey was a portion of an investigation into a county's facilities. It was primarily concerned with (1) the amount of usage of the jail facility, (2) the types of offenders being jailed, (3) alternatives to jail incarceration, and (4) the needs for a new jail, both in physical plant and in personnel.

The information forms were filled out using records from the present jail, the police department, the district attorney's office, and court records.

Each survey form was given a unique identification number for the purpose of separating each person's record in the computer, and having the ability to access a specific record by number if changes or additions were necessary. The following survey form has been reproduced entirely for this first detailed study. All questions will have a code for a data-entry person to use if the question is not answered or information is unknown. This can be any number that is not a possible response to a question, but usually 0s or 9s are used to be consistent. Question 15, number of previous felony arrests, for example, would use "99" if that question was unanswered (unknown), and the data-entry person would insert these two numbers in the computer file records for those applicable surveys. It had been predetermined that two numbers ranging from 0 to 98 would be sufficient for this question. The larger numbers may sound extreme but it was possible that a number greater than 9 might be needed; thus, in the column format, we have legal values of 0 to 98 with 99 used for unknown. Various other questions will make use of the number 9

COUNTY JAIL SURVEY

ID Number (assigned by surveyor) ____

1. Year of birth ____
2. Date of arrest (mo/day/year) ____
3. Time booked (hour 01–24) ____
4. Sex (circle one)
   1. Male 2. Female
5. Race (circle one)
   1. White 2. Black 3. Other
6. Residence (circle one)
   1. In county
   2. Other county in state
   3. Out of state
   4. Other
7. Employment (circle one)
   1. Employed
   2. Unemployed
   3. Student
   4. Military
   5. Retired
8. Category of charge (circle one)
   1. Felony 2. Misdemeanor 3. Civil
9. Charge (circle one)
   1. Aggravated assault
   2. Arson
   3. Burglary
   4. Larceny
   5. Motor vehicle theft
   6. Murder
   7. Rape
   8. Robbery
   9. Curfew-loitering
   10. Dangerous drugs
   11. DUI
   12. Other traffic
   13. Drunkenness
   14. Embezzlement
   15. Family offense
   16. Forgery
   17. Fraud
   18. Gambling
   19. Liquor laws
   20. Natural resources violation
   21. Other offenses
   22. Prostitution
   23. Public disturbance
   24. Runaway
   25. Sex offenses
   26. Simple assault
   27. Stolen property
   28. Vagrancy
   29. Vandalism
   30. Warrants
   31. Weapons
10. Arresting agency (circle one)
    1. Local police
    2. State Patrol
    3. FBI
    4. State BI
    5. Other
11. Type of release (circle one)
    1. R.O.R.
    2. Surety bond
    3. Property bond
    4. Cash bond
    5. Transferred
    6. Not released

12. Date of release (mo/day/year) __________
13. Time of release (hour 01–24) __________
14. Amount of bond (0–999998) __________
15. Number of previous felony arrests (0–98) __________
16. Number of previous misdemeanor arrests (0–98) __________
17. Disposition of charge (circle one)
    1. Filed 2. Dropped 3. Charge lowered
18. Date of arraignment (mo/day/year) __________
19. Date of disposition (mo/day/year) __________
20. Date of sentence (mo/day/year) __________
21. Number of passes (0–98) __________
22. Final disposition of charge (circle one)

1. Guilty	4. Incompetent	7. Bond forfeiture
2. Not guilty	5. Nolle Pros	8. Other (dismissed, mistrial)
3. Acquitted	6. Abandoned	

23. Sentence (circle one each if needed)[1]

Sentence 1	Sentence 2	Sentence 3
1. less than 1 year	1.	1.
2. 1–2 years	2.	2.
3. 2.1–5 years	3.	3.
4. 5.1–10 years	4.	4.
5. More than 10 years	5.	5.
6. Death	6.	6.
7. Fine	7.	7.
8. Sentence suspended	8.	8.
9. Probation	9.	9.

24. Amount of fine (circle one)

1. \$100 or less	4. \$1,001–\$5,000
2. \$101–\$300	5. Over \$5,000
3. \$301–\$1,000	6. None

25. Failure to appear this release (circle one)
    1. Yes 2. No 3. N/A
26. Rearrested while on release (circle one)
    1. Yes 2. No 3. N/A

Sentences 2 and 3 are for the cases in which a person received more than one sentence for a charge, e.g., time plus a fine.

to distinguish between questions answered as "other" and questions where the answer is unknown (numbers 5, 6, 10). In most surveys or polls we want to know the number of questions to which there was no response and the number of people who did not respond to certain questions.

Notice that questions are simple and request all answers in numeric form. Whenever possible, provision is made to circle answers that can be expressed in the form of multiple-choice questions. These conditions are ideal for processing by computer and they save time for the people completing the survey, if no automatic method of data entry is available.

These forms were completed on 897 records; the data was entered into the computer by a data-entry person and stored for processing later. (If all data is numeric a data-entry operation is much faster.) The next step was to enter the questions and possible answers into the computer. The method for this depends upon the type of computer statistics package you have and how you want the results to look. It can be simple, for your use, or more detailed for publication. Next you enter the data locations (section 6.3), the names of the statistics you want, and the question numbers for which you want these statistics, plus any arithmetic you want performed, and the statistics wanted for these items; for example, age was calculated by the computer by subtracting year of birth, question 1, from present year. Year of birth was recorded instead of age. Any follow-up studies in subsequent years could always use the same method of calculating age, so this data item does not have to be changed if used later.

The statistics wanted on this survey should have been determined before the questions were developed. Of course, we want a frequency table on all questions to see how the data looks, and some graphs, if appropriate for publication. Most of our questions involve counts of how many fall into each category. Therefore, other than frequencies by table and graph, the cross-tabulation with chi-square is the only appropriate statistic. Those questions that request a number rather than a circle answer can be used with the other statistics covered in this book—those statistics that require continuous measurement data.

For ease of processing by computer, military time was used for time booked, question 3. For appearance, it was converted on the graph, Figure 7-1. As expected, the frequency of arrests was greatest from 11:00 PM to 2:00 AM with the mode of 80 at 2:00 AM. A me-

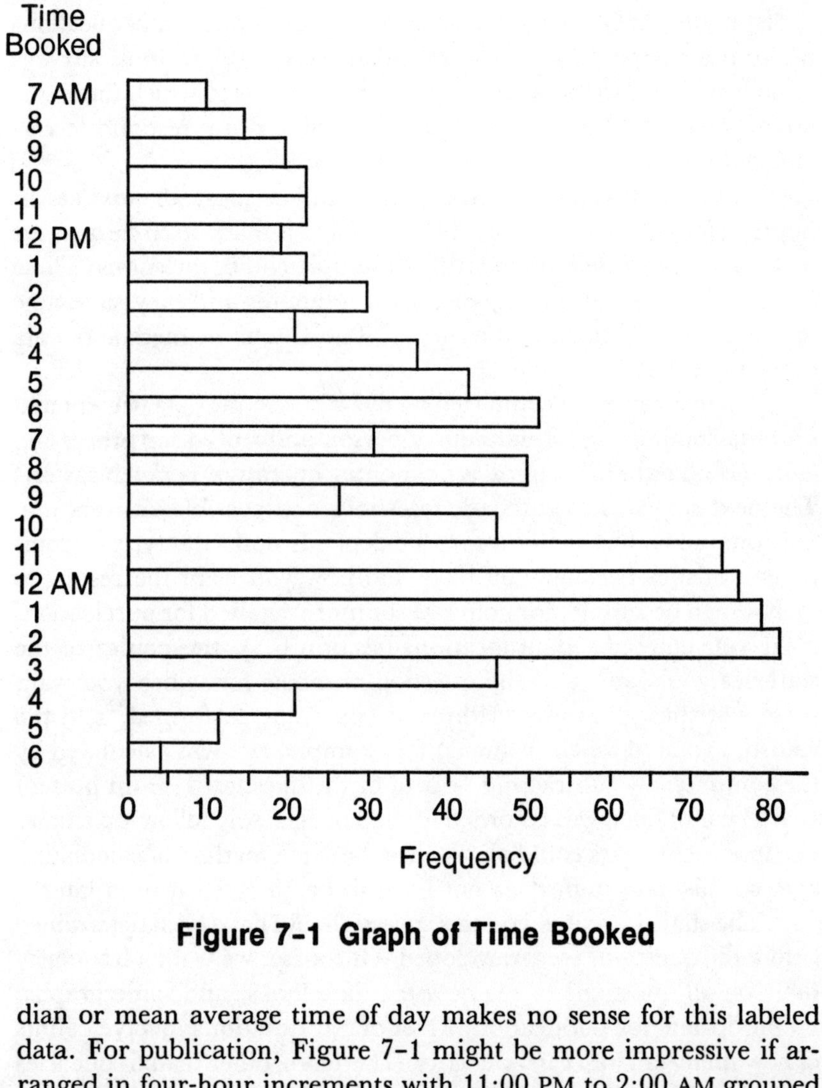

Figure 7–1 Graph of Time Booked

dian or mean average time of day makes no sense for this labeled data. For publication, Figure 7–1 might be more impressive if arranged in four-hour increments with 11:00 PM to 2:00 AM grouped together, then 3:00 AM to 6:00 AM, 7:00 AM to 10:00 AM, etc. For the purpose of this study, minutes between hours were not important. Therefore, the data gatherers recorded time to the closest hour. We did not have to overlap hours in the graph because there was no recorded data falling between 6:00 AM and 7:00 AM, for example. In grouping some data, care must be taken to avoid obscuring the conditions that should be noticed. The range for the groupings is your choice; the computer will automatically compress the bottom frequency line and numbers for placement on one page, as shown in Figure 7-2. The 11:00 PM to 2:00 AM time frame stands out as it should.

In Table 7-1, type of charge, using question 9, the relative frequency percent column is the column of interest. Almost half of the arrests were for DUI, response number 11. Include other traffic violations, response number 12, and we have over half, 54 percent (44.8 + 9.2), involving traffic. Cross-tabulations were also done to look at the frequencies in relation to other items, such as final disposition and sentence. However, the statistics were not reliable because some categories have a frequency of less than 5—arson, murder, rape, gambling, and weapons. As you see from Table 7-1, some response numbers had a frequency of 0, missing from table—items 9, 14, 19, 22, 24, 25, and 28. Again, the cumulative frequency column was not reproduced; there is no particular order for which we need to cumulate frequencies. The computer produced the table in response number order, 1 through 31 in question 9; if we had wanted to arrange the type of charge in some order, then this should have been the order of the possible responses on the original survey. In some cases, a definite order should be considered that would facilitate locating the proper number to circle when recording data (another example of

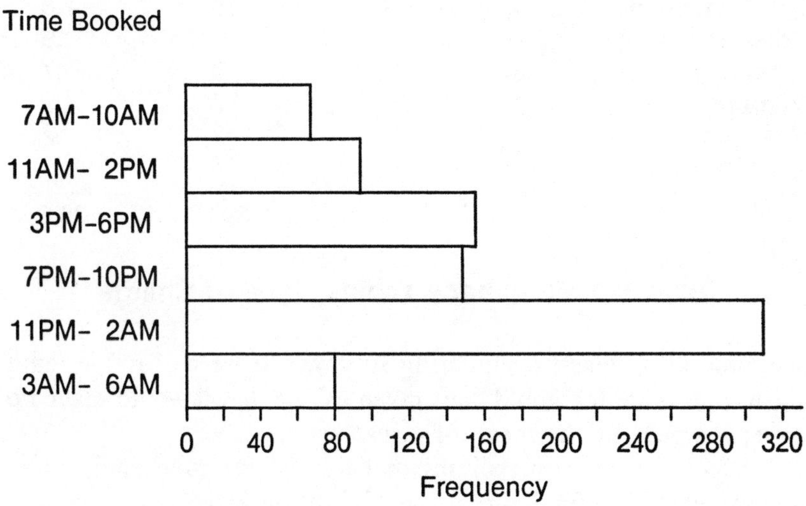

Figure 7-2 Graph of Time Booked, Grouped

Type of Charge	Response Number	Frequency	Relative Frequency Percent
Aggravated Assault	1	13	1.5
Arson	2	1	.1
Burglary	3	25	2.8
Larceny	4	41	4.6
Motor Vehicle Theft	5	15	1.7
Murder	6	2	.2
Rape	7	1	.1
Robbery	8	8	.9
Dangerous Drugs	10	62	7.0
DUI	11	398	44.8
Other Traffic Violations	12	82	9.2
Drunkenness	13	28	3.1
Family Offense	15	11	1.2
Forgery	16	8	.9
Fraud	17	10	1.1
Gambling	18	2	.2
Natural Resources Violations	20	10	1.1
Other Offenses	21	79	8.9
Public Disturbance	23	25	2.8
Simple Assault	26	16	1.8
Stolen Property	27	8	.9
Vandalism	29	6	.7
Warrants	30	36	4.0
Weapons	31	2	.2
Missing Data	99	8	
Total		897	100.0

Table 7-1 Frequency Table—Type of Charge

planning ahead when formulating surveys). If we had requested a mean, the computer would have given us one, but it would make no sense; it would be the mean of a response number.

A look at the percent column in Table 7-2 for question 6, shows almost half (100 − 53.8, or 13.5 + 32.7) of the arrests at this county jail were of persons living out of this county. Of these, a high percentage (32.7) lived out of state. This suggests a lot of activity caused

Legal Residence	Response Number	Frequency	Relative Frequency Percent	Cumulative Frequency Percent
In County	1	478	53.8	53.8
Other County in State	2	120	13.5	67.3
Out of State	3	290	32.7	100.0
Missing Data	9	9		
Total		897	100.0	

Table 7-2 Frequency Table—Residence

by tourists and transients. The missing data was not included in the percentages. With a simple instruction to the computer, the percentage could have been adjusted to account for these 9 persons.

Time spent in jail in hours was calculated by the computer from date and time booked, questions 2 and 3, and date and time of release, questions 12 and 13. Table 7-3 shows some results. The extremely high mean in relation to the median and mode, and the extremely high skewness, make the mean worthless as a description of this data. The 95 percent C.I. is also affected by this—the negative 376 hours is an impossible time frame. The median of 10.2, a better descriptor, tells us half of those arrested spent 10 hours or less in jail. The mode of 1 hour, meaning most people were detained for only 1 hour, may indicate a lot of unnecessary activity that could be alternatively handled, thereby saving time and manpower. This is also true in the figure for the median. The frequency table was also checked for any other modes. There were none. The minimum of 0 shows at least one person was detained less than one hour (actually less than one-

Mean	1010.4	S.D.	18135.3
Median	10.2	Skewness	25.4
Mode	1.0	Kurtosis	651.0
Minimum	0	C.V.%	1794.9
Maximum	464622	95% C.I.	-376 to 2396
Range	464622	Std. Error	693

Table 7-3 Time in Jail in Hours

half hour recorded to the nearest hour). For publication, the type of information that might be misunderstood should be denoted on the table, in this case, the fact that 0 means less than one-half hour.

The range, standard deviation (S.D.), and coefficient of variation (C.V.), show a lot of variation in time spent in jail. Although the variation was large, the large positive skewness tells us that only a few were on the upper end of time spent with a concentration on the lesser hours (section 3.1). This figure for skewness is extremely large compared to what is usually seen. A look at the full frequency table by hours showed the maximum hours spent in jail to be alone at the top end. There was a large gap between this extreme value of 53 years and the preceding figure of 2 years (converted from hours to years). One extreme data value, on either end, after a large gap in the data, is sometimes referred to as an *outlier.* In this particular case, it may have been wise to omit this one atypical value from the statistics. It is possible that this was a mistake in recording data originally, or that the data-entry person was distracted.

The extremely high positive kurtosis denotes an extreme peak in the data. From the frequency table this peak was found to be between 1 and 15 hours. The computer produced a frequency table for time in jail, but this particular one was not reproduced because of its length. For a report to the public on this item we could categorize the time spent as 0 to 12 hours, 13 to 24 hours, 25 to 48 hours, 49 to 72 hours—with the last category using the "more than" label, or we could have converted to weeks or months instead of hours. This can be done with instructions to the computer and a table or graph can then be produced. However, converting to weeks or months would obscure the predominant conditions of so many spending less than 1 day, or only 1 hour, in jail. The days spent in jail were also figured by the computer by dividing by 24. Again, since so many spent less than 1 day, the time in hours rather than in days is more useful as a decision tool.

The date of arrest, question 2, was converted by the computer to a Julian date and subsequently to a day of week. A cross-tabulation table was produced that compared day of week to category of charge, felony or misdemeanor or civil, using question 8. There were no charges recorded in the civil category. In Table 7-4, the top figure in each cell is the number of persons, the second figure is the row percent, the next figure is the column percent, and the bottom figure

Weekday	Category of Charge Felony	 Misdemeanor	Total
Sunday	17 12.1 9.8 2.0	124 87.9 18.7 14.8	141 16.9
Monday	28 33.7 16.1 3.3	55 66.3 8.3 6.6	83 9.9
Tuesday	21 26.9 12.1 2.5	57 73.1 8.6 6.8	78 9.3
Wednesday	23 24.0 13.2 2.8	73 76.0 11.0 8.7	96 11.5
Thursday	21 24.7 12.1 2.5	64 75.3 9.7 7.7	85 10.2
Friday	36 24.8 20.7 4.3	109 75.2 16.5 13.0	145 17.3
Saturday	28 13.5 16.1 3.3	180 86.5 27.2 21.5	208 24.9
Total	174 20.8	662 79.2	836 100.0

Missing Data = 61

DF = 6
Chi-square = 26.3
Significance = .000

Table 7-4 Cross-tabulation—Weekday by Category of Charge

is the total percent (section 5.3). The two top cells, for example, would be interpreted thus: Seventeen people were arrested on Sunday for felonies and 124 people were arrested on Sunday for misdemeanors; 12.1 percent of those arrested on Sunday were for felonies and 87.9 percent of those arrested on Sunday were for misdemeanors; 9.8 percent of the total felonies for the week were arrested on Sunday and 18.7 percent of the total misdemeanors for the week on Sunday; 2 percent of the total of all arrests were felonies on Sunday and 14.8 percent of the total of all arrests were misdemeanors on Sunday. The largest number of felony arrests (looking down the first column) was on Friday, 36 or 20.7 percent, and the largest number of misdemeanor arrests (looking down the second column) was on Saturday, 180 or 27.2 percent. The largest number of total arrests (looking down the outer column) was on Saturday, 208 or 24.9 percent. Looking at the other figures in Table 7-4, we see that most arrests were on Friday, Saturday, and Sunday, especially the misdemeanors. The felonies don't show quite as much difference—with more felonies on Monday than on Sunday, and an equal number on Monday and Saturday. The large chi-square and significance less than .05, in fact less than .001, denotes a relationship between weekday and category of charge.

The next statistics were obtained to determine the time lapse between arrest and final disposition (final hearing). To calculate this time lapse in days the computer used date of arrest, question 2, and date of disposition, question 19, by first converting months and days to Julian dates. Table 7-5 shows the results representing number of days. The median of 40.2 indicates about half stretched out over 40 days up to a maximum, for at least one person, of 586 days. The mean average time was 92½ days; the low positive skewness shows a few extreme cases on the upper end; however, there were no large gaps in the data, no outlier such as there was in Table 7-3, therefore the skewness is low positive. The low positive kurtosis denotes a slight peak or peaks in the number of days. The minimum of 0 shows at least one for less than one day. The range, S.D., and C.V. show a lot of variation in the time between arrest and final disposition. The 95 percent confidence interval (C.I.) indicates the span of days expected if the mean of 92½ days is representative in this county; that is, future dispositions can be expected to take, on the average, between 81½ and 103½ days from the time of arrest, if we want to predict

Mean	92.5	S.D.	115
Median	40.2	Skewness	1.7
Mode	1 and 39	Kurtosis	2.3
Minimum	0	C.V.%	124.3
Maximum	586	95% C.I.	81.5 to 103.5
Range	586	Std.Error	5.5

Table 7-5 Time between Arrest and Final Hearing in Days

the future from patterns of the past. The 95 percent confidence interval is the mean + or − 11; this is also + or − 2 standard errors (+2 times 5.5 or −2 times 5.5).

Statistics were also obtained for number of days between date of arrest and date of arraignment. As the interpretation of the statistics is the same, it was not reproduced.

Age, question 1, is usually a statistic of interest. A frequency table was produced and the main statistics relating to age are shown in Table 7-6. The youngest arrested was 13 (minimum) and the oldest was 71 (maximum). Although the range of 58 is a small number, it is a wide range when you are considering ages. Most of those arrested were 22, mode, and half of those arrested were less than 26, median of 25.9. The average age was 29.6. Positive skewness shows fewer in the older age groups and more in the younger age groups, but no outliers to inflate the skewness. Kurtosis shows a slight peak or peaks in the data. Variation is relatively large at 36.3 percent. The 95 percent C.I. predicts the average age in future data should be between 28.9 and 30.3. This is the mean + or − .7, or + or − 2 (.35), the standard error. The statistics in Table 7-6 have to be computed with the data ungrouped. If the data is grouped, it then becomes labeled

Mean	29.6	S.D.	10.7
Median	25.9	Skewness	1.3
Mode	22	Kurtosis	1.4
Minimum	13	C.V.%	36.3
Maximum	71	95% C.I.	28.9 to 30.3
Range	58	Std.Error	.35

Table 7-6 Age

data rather than continuous data. To get a concise picture of ages, it was grouped for Figure 7-3. Each group, except the first and last, shows a span of six years. The data was grouped simply by entering into the computer those figures in the column under age. The computer did the rest. Age was computed by year; if it were otherwise, our group would have been 18–24, 24–30, etc., to allow for months between years. There were very few under 18 and these would have been transferred to a juvenile facility. These few contributed very little to the mode of one hour spent in jail in Table 7-3.

The type of charge, question 9, was also regrouped into the broad categories of person offenses, property offenses, and other offenses, and compared with several items. The person offenses include aggravated assault, murder, rape, robbery, and simple assault. The property offenses include arson, burglary, larceny, motor vehicle theft, stolen property, and vandalism. The rest were included in the other offenses category. One of these comparisons is shown in Table 7-7

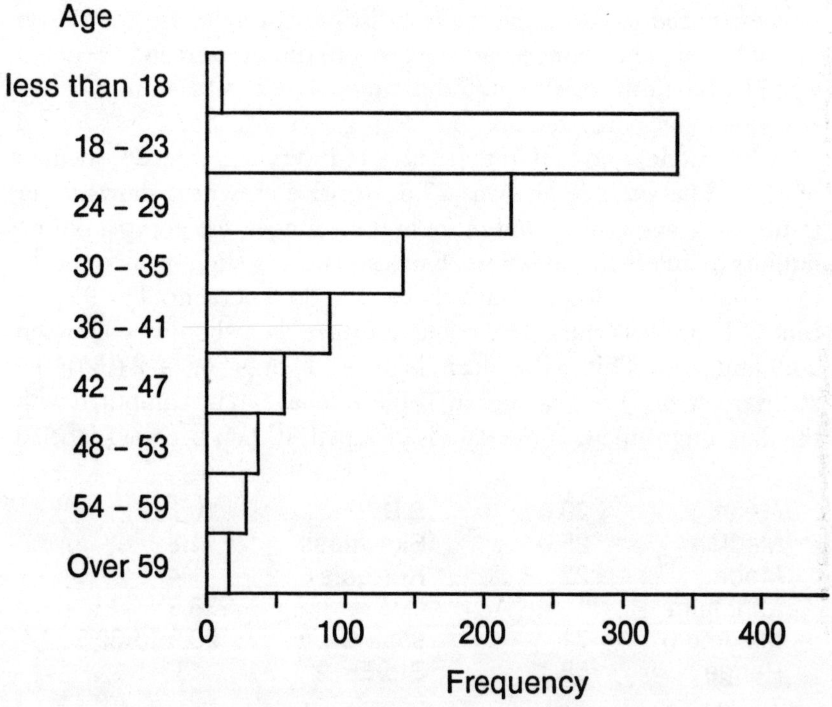

Figure 7-3 Graph of Age, Grouped

Sentence 1	Number of Previous Felony Arrests: 0	1 or More	Total
Received a Sentence	44 81.5 23.7 20.9	10 18.5 40.0 4.7	54 25.6
Did Not Receive a Sentence	142 90.4 76.3 67.3	15 9.6 60.0 7.1	157 74.4
Total	186 88.2	25 11.8	211 100.0

DF = 1
Chi-square = 2.29
Significance = .13

Table 7-7 Cross-tabulation—Type of Charge by Category of Other Offenses

for the other offenses category; after grouping we did a *pick* (section 6.3) of those in the other offenses category. Number of previous felony arrests, question 15, was also regrouped for this table. Sentence 1, question 23, was regrouped into those receiving a fine, suspended sentence, or probation as one group, and those receiving a sentence into the other group.

Although the figures in Table 7-7 appear to show a relationship, the chi-square is low and the significance is not less than .05, or even less than .10. There is much missing data with only a total of 211 out of the 897 cases recorded. This is not strong enough to claim a relationship between having a previous felony arrest and receiving or not receiving a sentence. After the regrouping we have at least five in each group, which was not true in Table 7-1. Interpreting the figures outside the cells we have a total of 54, or 25.6 percent, receiving a sentence; a total of 157, or 74.4 percent, not receiving a sentence; 186, or 88.2 percent, with no previous felony arrests; 25, or 11.8 percent, with previous felony arrests. Inside the cells we have 44 persons with no previous felony arrests who received a sentence and 10 with previous felony arrests who received a sentence; 142 with no previous felony arrests did not receive a sentence and 15 with previous felony arrests did not receive a sentence. The second set of figures in each cell totals 100 percent across and represents sentence or no sentence; the third set of figures totals 100 percent down and represents no previous felony arrests or some previous felony arrests; and all four of the bottom figures in each cell together total 100 percent, each figure representing one of the four groups into which the sample was divided.

Most of the remaining questions in the jail survey, not previously covered, are summarized in Table 7-8. It is simply a manual picking out of the largest frequencies from the many tables produced by the computer.

After reviewing the statistics, it was decided that a smaller jail than originally requested would serve the needs (by itself, a total number jailed does not tell us enough about usage). The statistics also provided some detail about what the new jail ought to look like and how it should function: cell requirements, program space, various programs needed. It was also decided that alternatives for immediate booking and release should be found.

Questions		Largest Frequency	
4.	Sex	91.1%	male
5.	Race	67.5%	white
7.	Employed	58.5%	employed
10.	Arresting agency	85.9%	local police
11.	Type of release	74.5%	property or cash bond
14.	Bond amount	$300.	mode and median
16.	Number of previous misdemeanor arrests	72.6%	none
17.	Disposition of charge	88.6%	filed
22.	Final disposition of charge	34.3% 32.4%	bond forfeiture guilty
24.	Amount of fine	64.3%	$300 or less

Table 7-8 Summary of Frequencies by Percent

2. PHYSIOLOGICAL MEASUREMENT WITH EXERCISE

A doctor requested that some statistics be produced on data he had gathered in conjunction with a triathlon. The triathlon is a strenuous contest combining swimming, biking, and running. The events consisted of a 1.2 mile swim followed by a 53.3 mile bicycle race and a 13.5 mile run. The doctor was interested in (1) the measurement of myoglobin before and after exercise, and (2) the effect of strenuous exercise on other physical or physiological measurements, such as weight, hemoglobin, lactic acid, blood pressure, pulse, temperature, and respiration.

Myoglobin is an iron-containing protein, similar to hemoglobin, found in muscle and serving as a reservoir for oxygen and carbon dioxide. *(Webster's New World Dictionary,* second college edition) However, the emphasis in this report is on the possible statistics, not medical terminology or techniques.

Table 7-9 lists some of the data gathered from each participant, by type of data and when recorded, to simplify our choices on what statistics can be used. Unlike the jail survey, most of this data is measurement data rather than count data. This indicates the following major statistics will be appropriate for measurement data: mean, median, and standard deviation; scatter diagram, correlation, regression; t-test and analysis of variance. These can be produced for the total of the items or for the items grouped; for example, statistics on finish time for all twenty-five subjects as a group, or statistics on finish time for those answering yes to a certain question as opposed to those answering no to the same question. The labeled data in Table 7-9 can employ those statistics based on counts: frequencies, modes, and cross-tabulations.

The comparison of some of the prerace and postrace measurements was particularly interesting. The pairs t-test is appropriate for this—the before-and-after analysis. Three of these are shown in Tables 7-10 through 7-12 and a more condensed version is in Table 7-13.

Here, and throughout this section, *prerace* signifies immediately before race and *postrace* signifies immediately after race. The number of cases does not equal twenty-five in all tables. This could mean the measurement was not taken in the haste and confusion of the race, the recorder failed to record it or recorded an obviously unreasonable

Labeled Data

Sex
Yes or No Questions
- Prior triathlon participation
- Ever had a major medical illness
- Now taking medication
- Minor illnesses past month
- Muscle trouble past month
- Urine discoloration

Rating Questions: none, mild, . . . , very severe
- Muscle trouble; start, during, after each event
- Illness or confusion after race
- Intensity of effort during each event

Measurement Data

Prerace Data
- Age
- Height
- Length of training in weeks
- Intensity of training for each event in hours per week
- Intensity of training for each event in total miles
- Longest continuous workout in hours
- Number of days of carbohydrate load

Prerace and Postrace Measurements
- Weight
- Hemoglobin
- Lactic acid
- Myoglobin
- Blood pressure
- Pulse
- Temperature
- Respiration

Postrace or During-Race Data
- Water or other liquid consumed during triathlon, in pints
- Swim time
- Bike time
- Run time
- Total finish time

Table 7-9 Types of Data for Triathlon

	Number of Cases	DF	Mean	Standard Deviation	Mean Difference	r	t	2-tail Probability
Prerace	22	21	155.0	27.4	4.2	.99	5.26	.000
Postrace			150.8	26.2				

Table 7-10 Pairs t-test—Weight

	Number of Cases	DF	Mean	Standard Deviation	Mean Difference	r	t	2-tail Probability
Prerace	22	21	65.4	8.1	-32.8	.34	-12.4	.000
Postrace			98.2	12.6				

Table 7-11 Pairs t-test—Pulse

number, or the person did not finish the race. For number of cases of twenty-two in Tables 7-10 and 7-11, valid prerace, or postrace (or both), measurements were not available for three people. In the case of weight, pulse, hemoglobin, and myoglobin, a "0" was used as a signal to denote missing data to be omitted, as "0" is an impossible measurement for these. If either prerace or postrace measurement was 0 for a person, that person was omitted for that item comparison. Of course, the computer doesn't know that 0 is impossible, and is to be ignored, unless we have included this in our list of instructions (section 6.3).

Interpretation of Table 7-10

1. Mean and mean difference: average weight loss of 4.2 pounds from prerace to postrace.
2. S.D.: very close from prerace to postrace, little change in variability around the mean.
3. r: correlation is positive, appropriate use of t-test; correlation

is high, weight loss was uniform throughout the group; prerace weight (x) and postrace weight (y) are very close to a straight line on a scatter diagram.

4. probability: less than .001, definite significant difference between prerace and postrace weight.
5. t value: positive because top mean is greater than bottom mean; prerace average weight was greater than postrace average weight.

Interpretation of Table 7–11

1. Mean and mean difference: average increase of 32.8 from prerace to postrace; the mean difference is negative only because the postrace mean is larger.
2. S.D.: pulse rates of the group after the race showed more variability around the mean.
3. r: correlation is positive, appropriate use of t-test, but r is relatively low; the increase in pulse rate from prerace to postrace was not uniform; prerace pulse (x) and postrace pulse (y) are not very close to a straight line; Figure 7–4 is included to show this low correlation on a scatter diagram. The "2" signifies that two people had a prerace measurement of 60 and a postrace measurement of 101.
4. probability: less than .001, definite significant difference between prerace and postrace pulse rate.
5. t value: negative because top mean is less than bottom mean; larger than that in Table 7–10 with the same DF (we can be even more confident in claiming a significant difference).

Interpretation of Table 7–12

1. Mean and mean difference: average increase of .2 from prerace to postrace.
2. S.D.: little change in variability around the mean.
3. r: correlation is positive and high; increase in hemoglobin measurement was uniform from prerace to postrace.
4. probability: greater than usual limits of .05 or .01, no significant difference between prerace and postrace hemoglobin measurements.
5. t value: close to 1, disregarding negative; no significant difference.

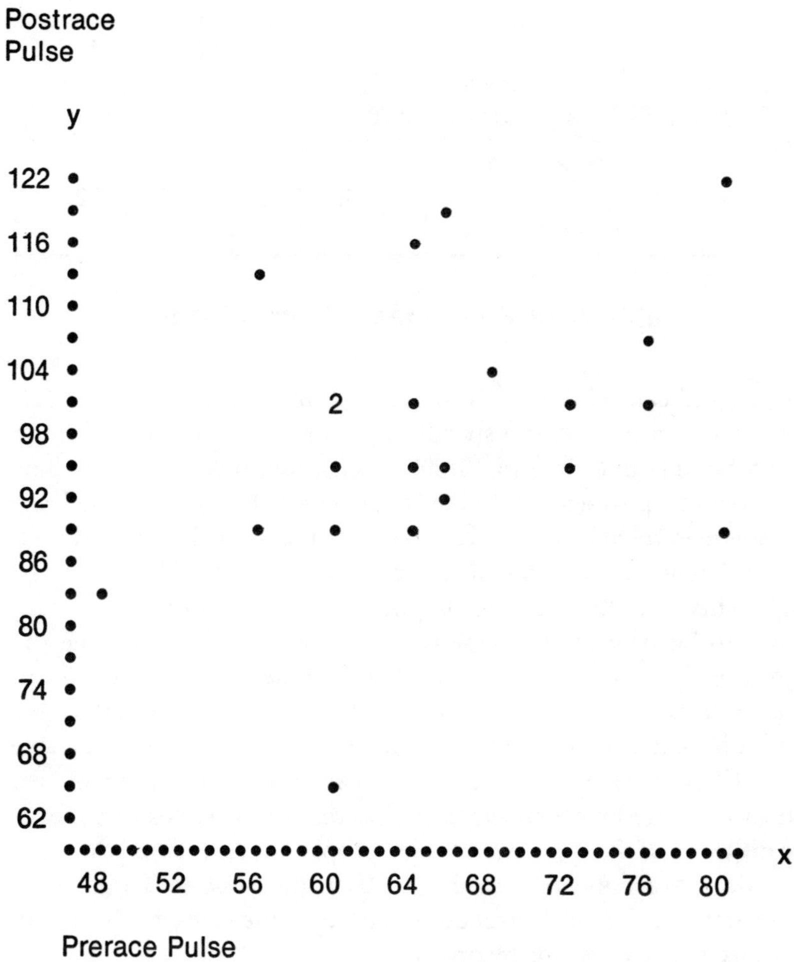

Figure 7-4 Scatter Diagram—Prerace Pulse and Postrace Pulse

As the doctor's main interest was in the measurement of myoglobin, the contestants were asked to return four times for this measurement and a few other measurements. The problem with these types of studies is that some people do not return, and the contestants are in uncontrolled environments in the interim. The comparisons are less reliable. In Table 7-13, twenty-four of the twenty-five had valid measurements for the prerace and postrace measurements. There

	Number of Cases	DF	Mean	Standard Deviation	Mean Difference	r	t	2-tail Probability
Prerace	24	23	15.2	1.2	−.2	.9	−1.3	.225
Postrace			15.4	1.4				

Table 7-12 Pairs t-test—Hemoglobin

was a significant difference in these measurements—in the last column less than .001. More specifically, our chances of being wrong would be only one time in 10,000 in claiming this difference. The next figure, postrace and one hour later shows no significant difference—probability of .842, greater than .05. We do have a significant difference again in the next one, postrace and four hours later—probability of .008, less than .05, or even .01. This difference continues to be true in the remainder of Table 7-13, as we see the significance, the increase of the t value, and the decrease of the mean in subsequent hours. Twenty-four hours later, the mean (103.7) is not yet back to prerace level (26.5), as shown in the first line of Table 7-13. Only seventeen people returned six hours later and eighteen returned twenty-four hours later. This is the reason all postrace means are different. If the same twenty-four people were measured throughout, all postrace means would have been the same in Table 7-13. The computer, by our instructions, omitted those cases in which both measurements were not recorded.

Event times were recorded in hours, minutes, and seconds. The computer cannot add 1:30 and 1:40, for example, and get 3:10; but, with a multiplier or divider entered by you, it can change these values to hours and fractions of hours, or minutes and fractions of minutes, or all seconds. This arithmetic can be performed by the computer on each contestant in the same way you would do it with pencil and paper—by multiplying, adding, and dividing. Hours × 3600 + minutes × 60 + seconds equals time in seconds. Divide by 60 and you have time in minutes and fraction of minutes. Divide by 3600 and you have time in hours and fractions thereof. A sophisticated statistics package may provide other ways of handling numbers that are not based on tens and hundreds. Event time statistics

	Number of Cases	Mean	t	2-tail Probability
Prerace	24	26.5	− 11.4	.000
Postrace		837.3		
Postrace	22	833.4	.2	.842
1 hour later		823.4		
Postrace	23	835.7	2.91	.008
4 hours later		675.7		
Postrace	17	890.6	5.81	.000
6 hours later		571.5		
Postrace	18	794.4	10.27	.000
24 hours later		103.7		

Table 7-13 Pairs t-test—Myoglobin

are shown in Tables 7-14 through 7-17. As every time was different, there were no modes. Skewness and kurtosis for all times were small—there were no extreme times and no concentration of data around certain times. All means and medians were relatively close within each event and total time. There was more variation in run time than any other, with a coefficient of variation of 23.5 percent in Table 7-16. Standard deviation is interpreted the same in all tables. For Table 7-17, total finish time, 68 percent of the contestants finished in the time of 5.86 + or − .83, or between 5.03 and 6.69 hours. The fastest person finished in 4.6 hours and the slowest in 7.57 hours. The computer could have processed this in seconds if we were interested in precision. This kind of precision was not needed for statistical comparison used in the t-test, correlation, and so forth.

Mean	31.14	S.D.	4.67
Median	30.32	Skewness	.97
Mode	none	Kurtosis	2.59
Minimum	22.07	C.V.%	15
Maximum	45.02	95% C.I.	29.17 to 33.11
Range	22.95	Std.Error	.98

Table 7-14 Swim Time in Minutes

Mean	182.84	S.D.	23.86
Median	175.93	Skewness	1.08
Mode	none	Kurtosis	.89
Minimum	153.83	C.V.%	13.1
Maximum	244.93	95% C.I.	172.52 to 193.15
Range	91.1	Std.Error	5.15

Table 7-15 Bike Time in Minutes

Mean	136.26	S.D.	32.02
Median	128.13	Skewness	.86
Mode	none	Kurtosis	– .14
Minimum	94.48	C.V.%	23.5
Maximum	208.23	95% C.I.	122.41 to 150.1
Range	113.75	Std.Error	6.92

Table 7-16 Run Time in Minutes

Mean	5.86	S.D.	.83
Median	5.72	Skewness	.59
Mode	none	Kurtosis	– .36
Minimum	4.6	C.V.%	14.08
Maximum	7.57	95% C.I.	5.51 to 6.21
Range	2.97	Std.Error	.17

Table 7-17 Total Finish Time in Hours

Correlations of event times were not of interest and not investigated. Some correlation should be expected. A person's poor performance in one event would have an effect on total time. Correlation or association between the physiological conditions or between the physiological and physical conditions were the items of interest. These were investigated for almost every possible pair of measurement data that was a reasonable comparison. An example of an unreasonable comparison is height versus age, which would be of no statistical value for this study. Very few measurements were found to be associated, even those that we might expect, such as, amount of training for the races and finish times, or age and finish times.

Some of the items of primary interest to the doctor for linear correlation were (1) the periodic measurements of myoglobin with pulse, (2) myoglobin with temperature, (3) myoglobin with lactic acid, (4) myoglobin with finish times, (5) myoglobin with intensity of training, (6) myoglobin with amount of liquid consumed during races, and (7) myoglobin with age.

None of these resulted in a high correlation. The highest correlation among items of lesser interest was found between prerace weight and postrace hemoglobin measurements. Correlation of .73 in Figure 7–5 was not high enough to make any definite claims without further study. If these could be shown to be associated under certain conditions, we could predict hemoglobin after exercise if we knew weight. For example only, without making any medical or statistical claims about an association, we will go through the use of the formula given in section 4.4. Hemoglobin is the dependent variable on the vertical (y) axis and weight is the independent variable on the horizontal (x) axis. You always know the independent variable is the one occurring first, prerace weight in this case, and the dependent variable is the second one in time, postrace hemoglobin. The formula is y = a + bx; y is the predicted hemoglobin, a and b were produced by the computer in Figure 7–5, and x is any weight. Calculating hemoglobin for 175 pounds we have

$$
\begin{aligned}
y &= a + bx \\
y &= 9.5 + .04(175) \\
y &= 9.5 + 7 \\
y &= 16.5.
\end{aligned}
$$

Y will not always be 16.5 when x is 175, but the standard error of estimate predicts that 68 percent of the time y will be 16.5 + or − .97, or between 15.53 and 17.47. R squared of .54 (.73 times .73 rounded) in Figure 7-5 means 54 percent of the variation in postrace hemoglobin can be attributed to prerace weight. Table 7-18 is the set of measurements that are plotted in Figure 7-5.

Some of the items of primary interest for significant difference with t-test were (1) contestants split into two groups according to answers to prerace questions (labeled data in Table 7-9) with all myoglobin measurements, and (2) myoglobin with sex grouped.

Tables 7-10 through 7-13 were examples of the pairs t-test. Table 7-19 is an example of the groups t-test. According to the table, the difference in measurements is significant between males and females,

Prerace Weight	Postrace Hemoglobin
116	11.4
124	13.9
125	13.6
127	12.6
130	15.3
131	14.4
132	15.7
134	15.5
141	15.3
148	15.5
149	15.9
150	15.5
154	14.9
155	17.0
157	16.3
158	15.8
163	15.2
168	15.4
184	15.9
191	15.6
191	16.6
192	16.4
223	17.7
2 missing	

Table 7-18 Prerace Weight and Postrace Hemoglobin

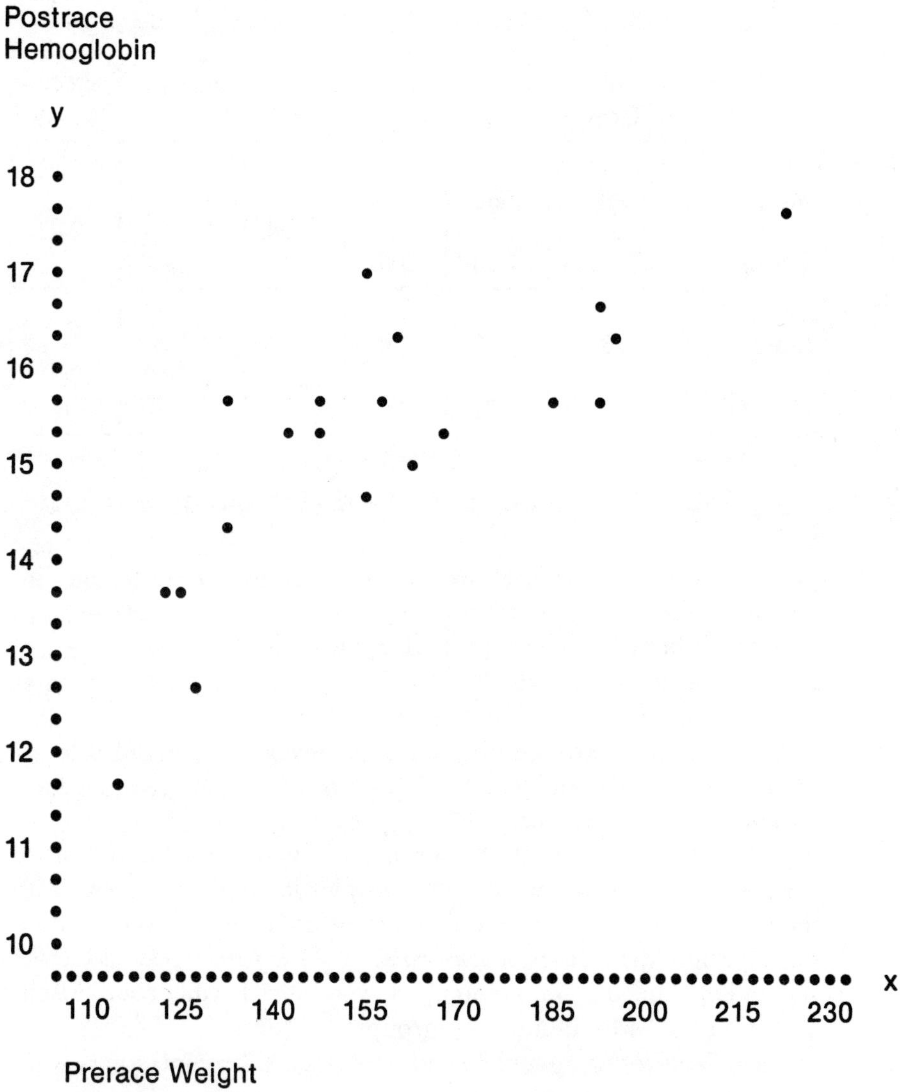

Correlation, r = .73 r squared = .54
Intercept, a = 9.5 Slope, b = .04
Standard error of estimate = .97 Significance = .000

Figure 7-5 Scatter Diagram—Prerace Weight and Postrace Hemoglobin

	Number in Group	Mean	Standard Deviation	F	t	2-tail Probability
Postrace						
Male	20	883	312	1.04	2.11	.046
Female	4	520	318			
24 Hours Later						
Male	16	113	54	29.75	5.38	.000
Female	2	31	10			

Table 7-19 Groups t-test—Myoglobin by Sex

probability less than .05. However, the small number of females invalidates any claims that could be made about the difference in myoglobin between the sexes, both postrace and twenty-four hours later. Notice that the large F value for data twenty-four hours later tells us that we do not have a comparable set of data. This F value is a result of the extreme difference in means and standard deviations between males and females. For postrace we do have more comparable groups; F is close to 1.

The items of interest for significant difference with analysis of variance (section 5.3) were the periodic myoglobin measurements with the rating questions. The t-test cannot be used with the rating questions because the ratings of *none, mild, . . . , very severe* resulted in the separation of the data into more than two groups for that comparison, and the t-test is limited to two groups.

In Table 7-20, myoglobin measurements for contestants were compared as groups split according to muscle problems. The aim was to determine if there was a significant difference between the measurements of those having and those not having muscle problems during the past month or during the events. The difference in the groups having or not having muscle cramps the past month is not significant, .893 in Table 7-20. The difference in groups with varying degrees of muscle trouble after the swim event was significant, at least less than .10. We will call it the 7 percent level of significance,

Source	DF	Sum of Squares	Mean Square	F	Significance
Muscle cramps past month (yes/no)	1	16.7	16.7	.02	.893
Muscle trouble after swim (rated by severity)	4	14702	3675.5	4.4	.068
Interaction	2	13532	6766	8.1	.027
Error	5	4175	835		
Total	12	33289	2774		

Table 7-20 Analysis of Variance—Myoglobin 24 Hours after Race

and the 93 percent level of confidence. The combination of the two produced a significant difference, less than .05, shown in the interaction significance; the effect of muscle trouble upon myoglobin was not the same throughout all the separations of data into yes/no, then severity levels.

The degrees of freedom (DF) for muscle cramps past month is 1 because we have two groups, yes or no. The DF of 4 includes 5 groups of ratings, from none to very severe. The total DF of 12 indicates that we had 13 people for whom all measurements, answers, and ratings were recorded. The mean square is obtained by dividing the sum of squares by DF. F is obtained by dividing the mean square by the error mean square.

For those of you who have a particular interest in this subject, some of the other results are shown, without explanation, in Tables 7-21 through 7-23.

The next time an attempt is made to gather data under these conditions, it should be better organized. Some of the inhibiting circumstances can be foreseen. As in most statistical studies by computer, planning ahead is much more time-consuming than producing the statistics. No broad medical or statistical claims can be made about this study. We can say only what was true in this one case, and then only about those statistics for which most of the data was recorded. The total sample was too small, and there were relatively too few females for groups by sex. The volunteer participants could not be forced to continue with the follow-up measurements. Although mis-

Mean	32.4	S.D.	7.7
Median	30	Skewness	.77
Minimum	20	Kurtosis	−.02
Maximum	51	C.V.%	23.9
Range	31	95% C.I.	29.1 to 35.6
		Std.Error	1.6

Table 7-21 Age

	Prerace	Postrace	Difference from Prerace to Postrace
Mean	98.3	97.99	Drop in average
Median	98.4	97.6	Drop in median
Minimum	97.2	95.5	Drop in minimum
Maximum	99.4	101.6	Increase in maximum
Range	2.2	6.1	More variability
S.D.	0.6	1.5	More variability
C.V.%	0.6	1.6	More variability

Table 7-22 Temperature

	Prerace	Postrace	24 Hours Later
Mean	5.9	22.8	10.8
Median	5.8	21.9	10.1
Minimum	1	12	5
Maximum	16	39	17
Range	15	27	12
S.D.	3.9	7.2	3.4
C.V.%	65.5	31.5	31.5

Table 7-23 Lactic Acid

takes have been noted throughout and unusable statistics shown, it was an interesting and well-done *first* study. For some of the usable statistics (not included in this book), the young doctor received recognition with the publication of the results in a professional journal. It is hoped that he will pursue his research contributions, encompassing more data on more cases in a more controlled environment.

8

Government and Education

1. CITY PLANNING COMMISSION SURVEY

The Planning Commission seeks the views of residents and nonresident property owners on future development of the city. Your views will help us plan for growth patterns acceptable to a maximum number of citizens.

Please note that you need not sign the questionnaire, if you prefer anonymity. You may use the enclosed envelope for your response.

Please take the time to answer and return the form promptly. Your cooperation is sincerely requested. Thank you.

The previous statements were the opening remarks of a questionnaire that was part of a twenty-year projection for a small city. The aim in this portion was to obtain majority opinions and help determine what directions to take for planning and growth.

The planning commission first contacted an outside consulting firm to conduct this survey for them. The quoted price for this service was $32,000. Seeing no justification for spending this amount of money, they decided to undertake the study themselves. They knew very little about computers or about statistics, but they were intelligent people. With a little help from interested volunteers, they could learn, and they already knew the questions for which they wanted answers and majority opinions from residents. Much of the work was clerical. In addition, they regarded this survey as a worthwhile challenge, to which they were willing to devote extra time (and it was politically expedient). They had access to a computer department and computer personnel from whom they received advice about question formulation and arrangement of data for processing by computer. Many people do not think about this aspect until after the survey has been returned. Then manual changes often are necessary—changes that could have been avoided with some preplanning for computer use.

As usual in surveys of this type, a stamped return envelope was provided. As a public service, TV and radio stations publicized the questionnaire and encouraged people to complete it. It was also promoted by civic organizations. The results were to become public information; the final 37 percent return rate was very good.

The survey in section 7.1 was given in entirety. For this example, we will randomly list some of the questions, one at a time, and then provide at least one method of presenting the statistics.

The section, or area, of the city might have an influence on peoples' attitudes about directions to take for planning and growth. Therefore, a skeletal city map was included in the original survey and the first question was aimed at categorizing responses according to the area of the city in which the person resided or owned property.

---

1. Please look on the attached map and decide which of the eleven areas is where you live or own property. Then put the number of that area here: ____________ .

This is an acceptable numeric resonse. However, as there were only eleven choices for this question, it is suggested that these eleven numbers should have been included on the survey as a number to circle instead of a blank to fill in. The handwriting of some people is so poor that even a simple number between 0 and 9 is often impossible to read. We have to play a guessing game about what the respondent intended the number to be.

The second and third questions also were aimed at categorizing responses, one being by age group and the other by type of resident, such as, owner, renter, residential property, commercial property, or combination of any. An example, categorized by age, is shown later in this section.

---

5. What types of economic development (businesses) do you think should be encouraged or discouraged? (circle one number for each)

	Encourage	Discourage
Amusements and Recreation	1	2
Tourism (day visitors)	1	2
Tourism (vacationers)	1	2
Light Manufacturing	1	2
Heavy Manufacturing	1	2
Construction Companies	1	2
Retailing	1	2

Now, we have the recommended number to circle for the fifth question of this survey. The relative frequency percent column of a frequency distribution table is produced as a modified bar graph in Figure 8-1. The portion of the graph to the left of 0 is the percentage of those circling number 1 for "encourage" for each response and that to the right of 0 is the percentage of those circling number 2 for "discourage". Using the skeletal city map, the question of areas of the city recommended for these businesses was addressed in a later question on the survey.

The next question is aimed at determining present attitudes of city government toward some of the broad functions of government.

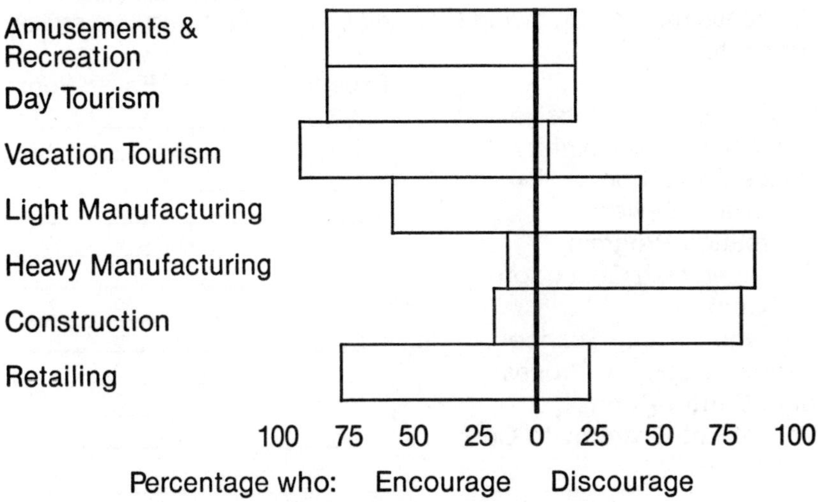

Figure 8-1 Graph of Desired Economic Development

9. Do you believe the political system encourages or discourages the following? (circle one number for each)

	Encourages	Discourages
Business	1	2
Tourism	1	2
Industry	1	2
Residential Growth (Low Density)	1	2
Residential Growth (High Density)	1	2
Cooperation with State, Federal Agencies	1	2
Media Coverage of Events	1	2

The results are shown in Figure 8-2, set up like Figure 8-1. The respondents consider the city to have the most emphasis upon publicity and cooperation, the last two in the graph. The city might claim that their favorable attitude toward these will cause other good things to follow. The politicians may not be encouraging tourism as much as the public would prefer, as indicated in Figure 8-1.

7. Please rate the following City Services by checking one space for each:

	Excellent	Good	Fair	Poor
Government Responsiveness	____	____	____	____
Financial Responsibility	____	____	____	____
Police Protection	____	____	____	____
Water and Sewer	____	____	____	____
Recreation Program	____	____	____	____
Garbage/Trash Collection	____	____	____	____
Pet Control	____	____	____	____
Fire and Rescue Service	____	____	____	____
Permit/Inspection Process	____	____	____	____
Sign Clutter Control	____	____	____	____
Vacant Lot Overgrowth Control	____	____	____	____

For question 7, the data was entered as 1 for "excellent," 2 for "good," 3 for "fair," and 4 for "poor," and only the percentages are produced in Table 8-1. Three levels of ratings may have been more appropriate for this; its main value lies in determining strong

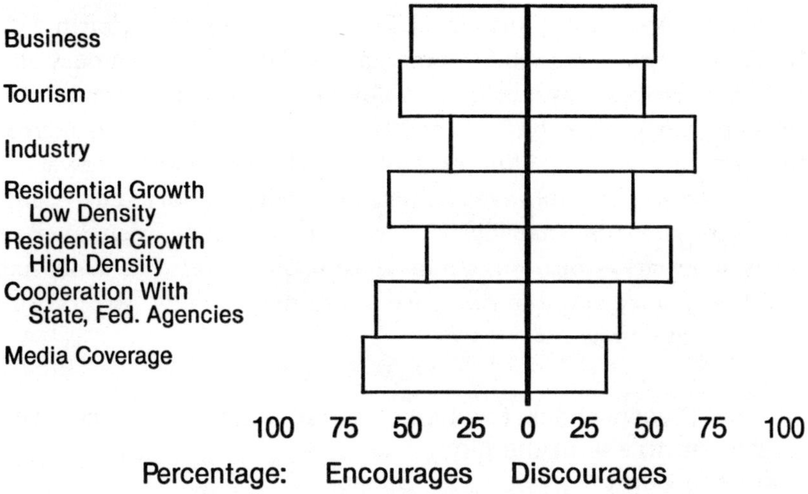

Figure 8-2 Graph of Present Political Attitude

points and weak points. As it is, the computer could group the 1s and 2s together and label this "good to excellent," then the 3s and 4s for "poor to fair" if we wanted to publicize a simpler table; we must avoid publishing misleading information that has been regrouped differently from the actual responses. The right way to request responses of this type is the way that is appropriate for your use. Often

	Excellent	Good	Fair	Poor
Government Responsiveness	20	45.2	25.9	8.8
Financial Responsibility	18.1	49.5	27.6	4.8
Police Protection	25.8	54	15.8	4.4
Water and Sewer	12.3	52	26.5	9.2
Recreation Program	4.7	33.8	40.5	21.1
Garbage/Trash Collection	21.7	48.9	21.4	8
Pet Control	7	31.5	28.2	33.3
Fire and Rescue Service	31.3	54.5	11.4	2.8
Permit/Inspection Process	9.4	47.6	28.1	14.8
Sign Clutter Control	4	31.3	34.5	30.4
Vacant Lot Overgrowth Control	3.3	17.1	34.5	45.1

Table 8-1 Rating of Present City Services by Percent

we want more separations for internal use. For example, birth date (or age) is often requested for some statistical uses as continuous data, and then grouped as labeled data for publication. It appears from the percentages in Table 8–1 that the city should improve its recreation program, method of pet control, sign clutter control, and vacant lot overgrowth control (over half checked fair or poor). The results of this question also were looked at as a comparison between different areas of the city, which was obtained using question 1. Questions like question 7 were asked in reference to utilities, ecology, streets, and zoning regulations.

---

14. Please check the types of additional businesses you would like to see in the city:

Shopping Center	_____	Banks	_____
Specialty Shops	_____	Dry Cleaners	_____
Department Stores	_____	Commercial Recreation	_____
Restaurants	_____	Service Stations	_____

The results of question 14 are shown as a standard bar graph in Figure 8–3. More restaurants may be wanted because the present ones are not good. This question also should be looked at as a comparison between different areas of the city. *Service stations* was checked by only 10 percent of the respondents. We can assume that 90 percent saw no need for any more service stations. If residents see more service stations appear in the city (permits issued), they may wonder why this survey was done. It would be wise to discourage new investors in this particular business, as well as the others with a low percentage desired. If the 10 percent of those wanting more service stations happened to reside in the same area of the city, then we could act accordingly. In place of additional businesses, the same question was asked about additional services, such as medical services, or a library.

The city has no city manager at present, so they asked for the feelings of the people about this in question 26.

---

26. Do you belive that the city should have a city manager? (circle one number)

1. yes 2. no

Figure 8–3 Graph of Additional Business Wanted

Figure 8–4 was produced using the appropriate row of percentages from the cells of a cross-tabulation table. The shaded area of the graph represents those replying yes and the unshaded area represents those replying no. It was interesting that as the ages increased, the larger the percentage of those not wanting a city manager. Forty-six percent of the total of the respondents replied yes to city manager and 54 percent replied no. If this created any indecision, the politicians might defer to the wishes of the older age groups (or to those where it would be politically beneficial).

The survey has implications beyond the simple responses to questions; for example, if vacant lot overgrowth control is so poor, maybe the personnel responsible for this should be investigated, or future budget priorities readjusted. This item may have received more criticism because it is something that is always in plain sight; it cannot be hidden from any segment of the public.

Other questions had the respondents rank priorities for physical public improvements, such as street surfacing and lighting; questions were also asked about support for certain social and economic improvements, such as historic site preservation.

The planning commission and city government were to formal-

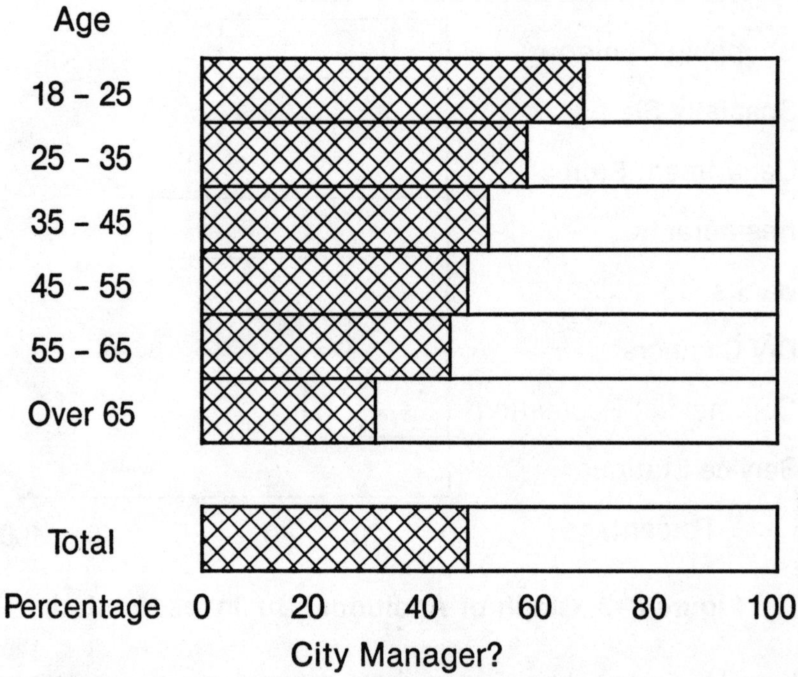

Figure 8-4 Graph by Age Groups: Request for City Manager—Shaded Portion is Percentage of Yes Responses

ly accept the results of this survey; it was not meant to be law, but simply a plan to influence every decision made about the use of land. Decisions that deviated from this plan would be obvious—and probably politically hazardous.

2. PREDICTING SUCCESS FROM GRADE-POINT AVERAGES

The theory behind this study was that success in one college major tends to be associated with success in certain other subject areas. As this can apply to any subjects, we will refer to this one college subject as the major. Our aim, then, is to (1) determine which subjects appear to contribute the most to a student's chosen major—which provide the best preparation; and (2) predict success or failure in a chosen major.

We will choose the major of physical therapy for this example. On the computer, along with other pertinent information, we have all the students' grades for each subject they have taken and their grade-point averages as they progressed in their college careers. In this case, grade-point average refers to mean average grades obtained after assigning a point system to grades of A, B, C, D, and F: A = 4, B = 3, C = 2, D = 1, and F = 0. The units of comparison were students' grade-point averages (GPA) in various subject areas, their overall GPA before they officially started a major of physical therapy in the third year, and their GPA in physical therapy at graduation. Using data from past graduates, for whom we had these measurements for comparison, we wanted to predict success or failure in this chosen major by knowing the results of the efforts of present students soon enough to prevent an ultimate failure.

For each student used in the initial study, we had measurements similar to those in Table 8–2. The table contains actual data for one of the students who was included in the statistics. All data is continuous over a limited range. For simplicity, the number of students' averages used is limited. The final GPA in physical therapy is what we are trying to predict. Therefore, it is the dependent variable. Using the scatter diagram and correlation on two continuous variables, we wanted to see if there was some relationship that would permit us to use one of the other measurements as a predictor of success in physical therapy. Enough of these are shown in Table 8–3 to review

GPA in PHYSICAL THERAPY at graduation	3.2
GPA in PHYSICAL THERAPY lab courses	3.4
GPA in PHYSICAL THERAPY lecture courses	3.2
Overall GPA after two years of college	2.7
GPA in all humanities courses	4.0
GPA in all English courses	3.0
GPA in all social science courses	3.1
GPA in all math courses	4.0
GPA in all chemistry courses	3.5
GPA in all physics courses	4.0
GPA in all biology courses	2.8
College entrance exam score	1150

Table 8–2 Grade-Point Average Record for One Student

Independent variable	Correlation	Significance
College entrance exam	.12	.253
Overall GPA after first two years	.29	.021
GPA in English	.25	.045
GPA in social sciences	−.09	.280
GPA in math	.30	.026
GPA in chemistry	.30	.025
GPA in biology	.31	.016

Table 8-3 Correlation with GPA in Physical Therapy at Graduation

the interpretations of correlation covered in Chapter 4. Although not good, the best predictor appears to be success in biology, having the highest correlation (r = .31) and the lowest significance (.016). Both math and chemistry are very close behind with r = .30. The college entrance exam has very little correlation (.12) with this particular major and the social science areas, correlation of − .09, may contribute to a well-rounded education, but not to success in this major for this selected group. Alone, none of these is sufficient for accurate predictions. All we can assume is that a student who has performed poorly in biology, chemistry, and math *may* need extra help or extra time, or perform better in a different major. We cannot add these three together and assume we have a good predictor.

The scatter diagram for biology GPA and physcial therapy major GPA at graduation is reproduced in Figure 8-5 with one pair of numbers plotted for each student. The student's averages from Table 8-2 would include a 3.2 on the vertical axis and a 2.8 on the horizontal axis, the first and last GPAs listed in the table. Correlation is positive, but not very strong. The data points are rather spread out on the graph. The range for correlation is from −1 to +1. For positive correlation, greater than 0, the straight line we try to approximate slopes upward from left to right; as one value increases, the other increases. In Figure 8-5, this is true in some cases but not enough to have a high positive correlation. The asterisks show this approximate straight line with some values lined in to emphasize the reason for the low correlation.

The value of y when x is 0 is the intercept, 2.4 (the asterisk on the y-axis); and the value of the slope, .2, indicates that for every

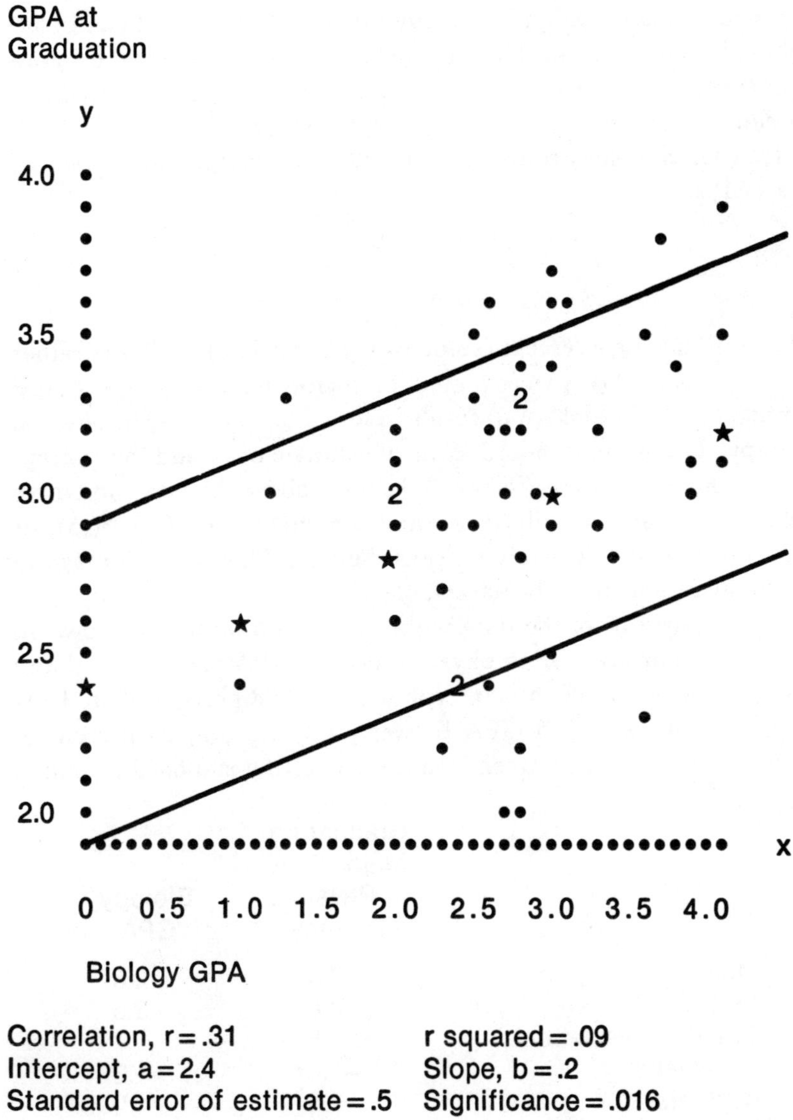

Figure 8-5 Scatter Diagram—Biology GPA with Physical Therapy Major GPA

change of one unit in x, y will change by .2 of a unit. R squared of .09 means only 9 percent of the variation in major GPA can be attributed to the variation in biology GPA and 91 percent can be attributed to other factors. The intercept and slope are used in the regression (prediction) formula, $y = a + bx$. If we use this to predict a student's average at graduation after the student has made a GPA of 1, or D, in biology courses, the result of predicting future y values if x is 1, is

$$y = a + bx$$
$$y = 2.4 + .2(1)$$
$$y = 2.6$$

The constant, or intercept (value of y when x is 0), of 2.4 is rather high. If biology GPA were a good predictor, we would expect an F average, or 0, in biology to result in less than C, a 2.4, in physical therapy. Using the standard error of estimate of .5 and the concept of division into six parts (section 3.3), three above the mean and three below the mean, we will have some students with $2.6 - 3(.5)$, or $2.6 - 1.5 = 1.1$, a barely passing average. This would not assure much success in the job market.

Comparisons of the data within these two items are shown in Table 8–4. For graduating physical therapy GPA the mean of 3 and standard deviation of .5 is interpreted thus: 68 percent of students should graduate with a GPA between 2.5 and 3.5, 95 percent of students should fall between 2 and 4 (twice the standard deviation

	Graduating Major GPA in Physical Therapy	Biology GPA
Mean	3.0	2.8
Median	3.0	2.8
Standard Deviation	.5	.7
Skewness	– .5	– .3
Kurtosis	– .4	.2
Range	1.9	3.0
Minimum	2.0	1.0
Maximum	3.9	4.0
C.V. %	16	25

Table 8–4 Descriptive Statistics of Two Sets

above and below the mean), and 99 percent should fall between 1.5 and 4.5 (three times the standard deviation above and below the mean). As a GPA of 4 is the highest, we should use 1.5 to 4. The mean and median are the same for both data sets and skewness and kurtosis are close to 0, which denotes fairly balanced sets of data, a normal distribution. If we compare one set against the other, the standard deviation, range, and coefficient of variation show more variability in the biology grades than in the physical therapy grade-point averages. We have at least one D average and one A average in biology, minimum of 1 and maximum of 4.

Although we found no one subject to be a good predictor, we can use these statistics to compare later groups of students. This set showed a mean of 2.8 and a standard deviation of .7 for biology. If later groups of students show a mean of 3.2 and a standard deviation of .4, for example, we know we have an exceptional group—and we know it from statistics, not just from our own perception.

Similar statistics were produced, based on how many courses were taken by students in each subject area. We were trying to determine if success in the physical therapy major was related to the total number of courses in each of the other subject areas; that is, if more exposure to certain subjects, regardless of grades, was a help. The correlations obtained were no better than those obtained for grades in the different subjects.

The determination and comparison of the simple statistics in this section led to the use of more advanced statistics not covered in this book. A combination of several subjects was found to be a good predictor of success in this major. The additional statistic used is called multiple regression. The concept is the same as that for simple linear regression. The main difference is the regression equation used in place of $y = a + bx$. Most statistics books will explain multiple regression for those who need it, and a short session with a statistician would be beneficial.

For comparison, the next section has a similar example with the results being more useful as a predictor.

3. PREDICTING SUCCESS FROM STANDARDIZED TESTS

In this example we are concentrating on a series of entrance exams or standardized tests to (1) predict success (or failure) in different subjects, and (2) track successful or unsuccessful students in a subject. The subject in question becomes the dependent variable and the standardized test is the independent variable. If the test is a good predictor, we can expect a high correlation between scores on the test and grade-point average in the subject. If not, if there is little positive correlation or any negative correlation, then the test is no good for this particular use. The grade-point average used is the same as that in section 8.2, and, as before, we have data from past students to compare their test scores with their GPA.

One of these subject areas, social sciences, is shown in Figure 8–6. Correlation of .73 is relatively high, indicating some relationship between entrance exam scores and GPA in social sciences. The figure for significance tells us that the chance of being wrong in claiming this relationship of .73 correlation is only 1 in 10,000. R squared of .54 means that 54 percent of the variation in GPA can be attributed to the variation in test scores and 46 percent can be attributed to other factors; for example, some students may be poor test takers on standardized tests. And students received their grades in social sciences from different teachers, in this case, an uncontrollable outside influence. The intercept of −.3 is more realistic than the one of 2.4 in the last section: if this test score is a good predictor, we can expect a low GPA to follow a test score of 0 (a small value for y if x is 0). The graph in Figure 8–6 does not include the possibility of 0 for test score on the x-axis, with the associated predicted −.3 for GPA on the y-axis, as there are no scores down to that range. The slope indicates a predicted rise of .04 in GPA for each one-unit rise in test score. The standard error of estimate, .4, indicates less expected error than in the example in section 8.2 in which the error was .5. This does not appear to be much difference, but we have to consider that this .1 difference is relative to the size of the original values, only a small range of 0 to 4 for GPA. Again, the asterisks are used to show the regression line along with the diagonal lines emphasizing less spread of data points than in the last example.

We can now use the regression equation and have the computer produce a list of students not expected to be successful in social sciences. The selected list is much more efficient, assuming the tests have been computer graded with grades entered into the students' records automatically. On the other hand, we can get a list of expected high grades for possible scholarships or course exemptions. A disadvan-

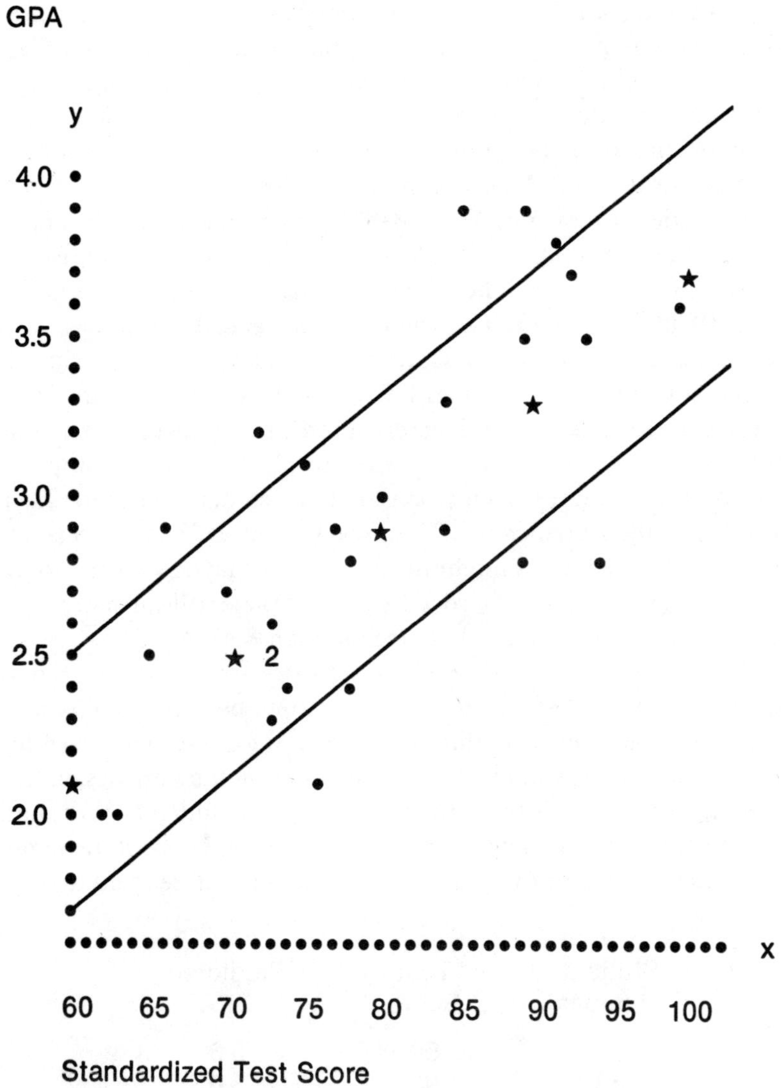

Correlation, r = .73
Intercept, a = −.3
Standard error of estimate = .4

R squared = .54
Slope, b = .04
Significance = .000

Figure 8-6 Scatter Diagram—Test Score with GPA

tage is that we have to work with a list by student identification numbers, as some statistics programs do not have the feature that produces lists of names. However, using code matching between files, names can be obtained from the students' numbers using other computer programs that are a part of the student computer file records system (section 6.1). The students' records are saved in the computer in alphabetical order. A portion of the results of this type of pick for higher grades is shown in Table 8–5. The predicted GPA has been calculated by the computer, student by student, using the regression equation, $y = a + bx$. The value for a was $-.3$, and the value for b was .04 in Figure 8–6. The value for the variable x changes with each student, and is the test score, which was computer produced, for that student. The predicted GPA was the calculated value of y for each student. If we are sincerely interested in the success of our students, we can also use these types of results for all of the students and have the computer produce a list of those students who later have fallen below their predicted GPA in social sciences. Some special attention to these students might be helpful—or maybe we need to investigate the work of their social science teachers if an inordinate number have the same teachers (see section 8.4).

The mean and other descriptive statistics are not reproduced as they were in Table 8–4. These two sets cannot be compared as easily. The possible values are different, 0 to 100 for tests and 0 to 4 for GPA. Its usefulness would be seen in comparing the statistics on this group of students with the same on subsequent students. We could see if later groups of students were better, or worse, or about the same. (Of course, the statistics do not tell us what our subsequent actions

Student Number	Test Score	Predicted GPA
26	90	3.3
32	98	3.6
27	89	3.3
61	93	3.4
16	87	3.2
25	86	3.1
89	95	3.5

Table 8–5 Predicted High GPA in Social Sciences

should be when educating poorer students or better students.) If we had a need for comparable sets of data, we could have the computer change both sets to z-scores using the formula in section 3.3 and the COMPUTE statement in section 6.3, or we could have the computer multiply each GPA by 25 before doing the statistics. This would result in both measurements being in the same 0-to-100 range.

4. COMPARING TEACHERS

The word *comparing* rather than *evaluating* has been used to avoid the implication that teachers can be judged on the basis of one statistical report.

Among students, a favorite topic of conversation is the difficulty or ease of getting a good grade from Miss R., or from Mr. S. "Try not to get Mr. X. You won't pass." A student will ask a teacher, "Why did you give me an F?" and the teacher will answer, "I didn't *give* it to you; you *earned* it." Evaluating teachers by the grades given, the method sometimes used by students, can be considered along with other ways of evaluating teachers. Statistically, we can determine the reasonableness of grades; were they earned or given? We will look at statistics based on the group of students assigned to each teacher to help us determine if (1) teachers appear to be grading as objectively as possible, and (2) their grading systems appear to contain the usual measurements.

For simplicity, we will use a pass/fail example rather than A, B, C, D, F. We can use this by a simple command to the computer to change the grade: A, B, C, D = P (for passed). This is a temporary change for obtaining the statistics. It does not affect a student's record of grades that were saved originally in the computer file records. We now have two rows for pass and fail in Table 8–6 instead of five for A, B, C, D, F. The computer printed the failed group first, in ascending alphabetical order. This table shows the number and percent of students passed and the number and percent of students failed for five teachers. Of course, these teachers are teaching comparable groups. We would not compare teachers of freshmen against teachers of seniors, for example.

In Table 8–6, we are interested in the first and third rows in particular (the counts of students and the percentage failed and passed by each teacher to see if any teachers passed too many or too few

	Teacher Identification Number 1	2	3	4	5	Total
Students Failed	33 22.9 30.3 4.7	49 34.0 30.2 7.0	30 20.8 21.3 4.3	12 8.3 9.0 1.7	20 13.9 12.6 2.8	144 20.4
Students Passed	76 13.5 69.7 10.8	113 20.1 69.8 16.0	111 19.8 78.7 15.7	122 21.7 91.0 17.3	139 24.8 87.4 19.7	561 79.6
Total	109 15.5	162 23.0	141 20.0	134 19.0	159 22.6	705 100.0

DF = 4
Chi-square = 33.1
Significance = .0000

Table 8-6 Cross-tabulation—Pass/Fail by Teacher

in comparison to the other teachers): teacher 1 has failed 33, 30.3 percent, and passed 76, 69.7 percent of the total of 109 students he or she taught this past term. Teacher 2 has failed and passed about the same percentage (30.2 and 69.8) as teacher 1. Teacher 4 has failed 12, 9 percent, and passed 122, 91 percent, of the 134 students he or she taught. The horizontal totals row at the bottom, other than total students, is read thus: teacher 1 had 15.5 percent of the total of 705 students, teacher 2 had 23 percent of the total of 705, etc. The vertical totals column to the right indicates that 144 students, 20.4 percent, failed and 561 students, 79.6 percent, passed. Some educators would call these last percentages about normal for a pass/fail rate. The figures for teacher 3 are very close to these total percentages with 21.3 percent failed and 78.7 percent passed. Actually the percentage each teacher passed or failed is the important figure, as all teachers did not have the same number of students; we cannot compare by count of students. The second and fourth figures in each cell have been ignored because we are not interested in the row percent or the percent of total within each cell. The row percent would tell us, for example, that of the students failed, 22.9 percent had teacher 1 and we would use this percent if we were comparing students instead of teachers.

The large chi-square and extremely low significance indicate that there is a relationship between grades and teachers—that the grades might depend on the teacher rather than on the students' efforts. Teacher 4 may have passed too many. These pass/fail percentages can be seen more clearly in the modified bar graph in Figure 8-7. The lengths of the areas for fail or pass from teacher to teacher are quite different, but we should not jump to conclusions. Maybe there are other factors. It is possible that some teachers actually had better students in their classes than other teachers had. This could occur with a random method of class assignment.

Our next step would be to have the computer group these students into so-called low achievers and higher achievers by looking at their

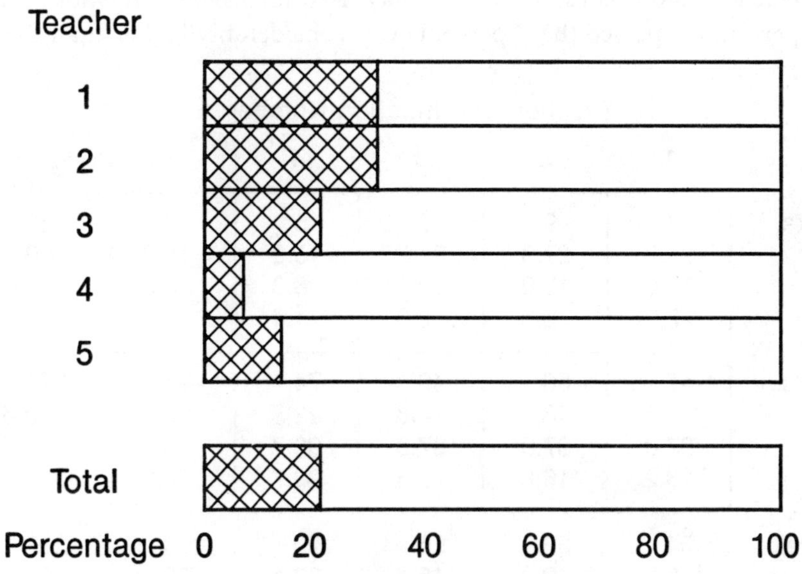

Figure 8-7 All Students: Graph by Percent Failed/Passed—Shaded Portion Is Percentage Failed

Teacher 3 is about normal compared to the total. Teacher 4 has failed the fewest.

grade-point averages (or some other standard measurement) up to this point in time. Then we would produce two more cross-tabulation tables to see if most of the failures were among the low achievers and most of the passes were among the high achievers. In Table 8–7 we now have a lower chi-square and higher significance. We should not claim much difference in teachers' grading. It appears that teacher 1 and teacher 2, whose failures showed a high percentage in Table 8–6, are not much out of proportion in Table 8–7 (reading across the third row of figures). Teachers 4 and 5 have a large percentage of the high achievers, in fact, half of them, 22.1% and 28%. Therefore, they should have the large percentage of passes. The vertical totals now show 9.2 percent failed and 90.8 percent passed. Figure 8–8 graphically portrays the closeness of the percentages. There is not a significant difference in the lengths of the bars for fail or pass.

In Table 8–8 we have a higher chi-square and lower significance. Among the low achievers there appears to be a relationship between teachers and grades. Teacher 4 may need further investigation; the percentage passed (87.3 percent) was considerably higher than that

	Teacher Identification Number					
	1	2	3	4	5	Total
Students Failed	7 21.2 13.0 2.0	9 27.3 13.0 2.5	7 21.2 12.7 2.0	5 15.2 6.3 1.4	5 15.2 5.0 1.4	33 9.2
Students Passed	47 14.5 87.0 13.2	60 18.5 87.0 16.8	48 14.8 87.3 13.4	74 22.8 93.7 20.7	95 29.3 95.0 26.6	324 90.8
Total	54 15.1	69 19.3	55 15.4	79 22.1	100 28.0	357 100.0

DF = 4
Chi-square = 5.8
Significance = .213

Table 8–7 Cross-tabulation High Achievers Pass/Fail

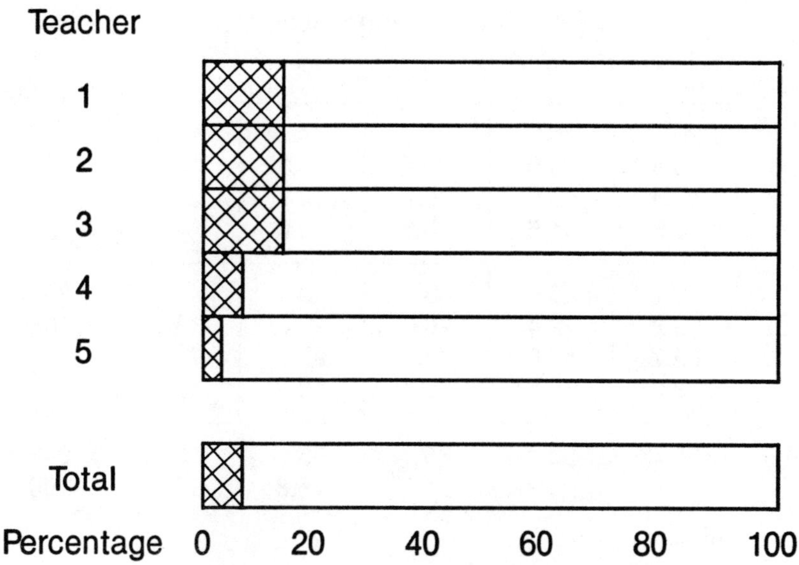

Figure 8-8 High Achievers: Graph by Percent Failed/Passed—Shaded Portion Is Percentage Failed

for the other teachers, and the percentage failed (12.7 percent) was considerably lower than the percentage for the other teachers. Again we can see a significant difference in the lengths of the bars in Figure 8-9. Of course, additional investigation by the principal (or supervisor) might show that teacher 4 is an excellent teacher for low achievers; that the students were not given grades but did, in fact, perform better for this teacher.

All educational institutions keeping records of students would have files looking somewhat like those in Table 8-9. To produce the statistics in Tables 8-6 through 8-8, we would have processed the teachers' grades for each course and updated the students' grade history files. For Table 8-6, all students with the same teacher number, year, and term in the students' grade history file would be grouped together by the computer. For Tables 8-7 and 8-8, the current grade-point average in the students' master file would determine how the students were regrouped. The two files would be matched by a common student identification number. Of course, we would have to enter a number within our computer list of instructions to include all those

	Teacher Identification Number					
	1	2	3	4	5	Total
Students Failed	26 23.4 47.3 7.5	40 36.0 43.0 11.5	23 20.7 26.7 6.6	7 6.3 12.7 2.0	15 13.5 25.4 4.3	111 31.9
Students Passed	29 12.2 52.7 8.3	53 22.4 57.0 15.2	63 26.6 73.3 18.1	48 20.3 87.3 13.8	44 18.6 74.6 12.6	237 68.1
Total	55 15.8	93 26.7	86 24.7	55 15.8	59 17.0	348 100.0

DF = 4
Chi-square = 22.8
Significance = .0001

Table 8-8 Cross-tabulation—Low Achievers Pass/Fail

greater than a certain GPA for the higher group and all those less than a certain GPA for the lower group, and the computer would sort them out into these two groups. For an idea of the impressive speed of a computer, the 705 students were grouped and counted, percentages calculated, and the three tables produced (ready to be printed) in nine seconds. Of course, time needed is determined by the type of computer equipment used.

Students' Grade History	Students' Master File
Student Number	Student Number
Course Number	Name
Credit	Address
Title of Course	Phone
Year and Term	.
Teacher Number	.
Grade	Current GPA

Table 8-9 Students' Files

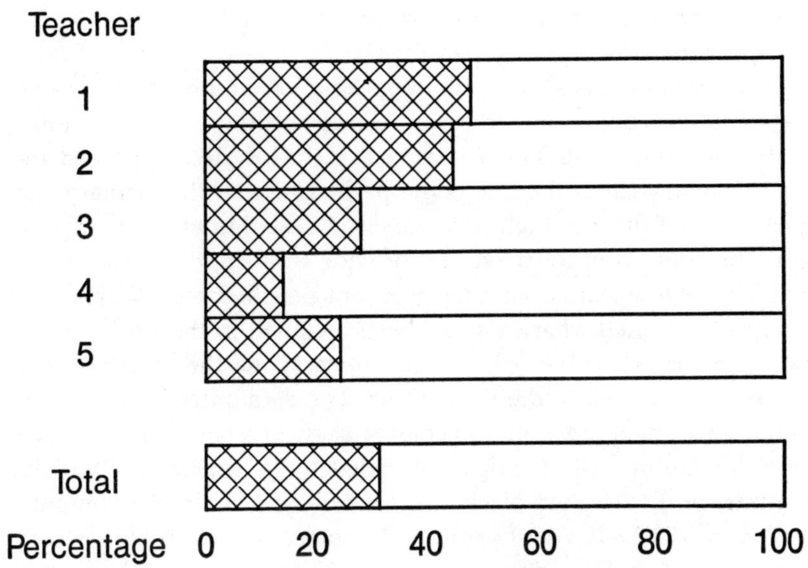

Figure 8-9 Low Achievers: Graph by Percent Failed/Passed—Shaded Portion Is Percentage Failed

5. PARKING HABITS

The last example in this chapter evolved into an absorbing class study; it touches on an observation that might be called one of human interest. As it seems that the phone always rings when you get in the shower, it seemed to one student that older or cheaper cars always parked too close to newer or more expensive cars. This student enlisted the aid of his psychology teacher and class to help him determine if his observations were correct. (If you ever see people out measuring distances between cars or recording data about cars, it is probably a class doing a statistical study; and, probably a psychology class. To get data on parking habits you watch people park and leave their cars, and you run up with your tape measure and data sheet.)

The cars included were those in which the drivers had a choice of parking spaces. This was done so that other factors could be controlled, for example, the space available in the area, a space closest to a store entry, a space in the shade, spaces reserved for smaller cars. The question is: Given a choice of parking spaces, (1) do people driving low-status cars tend to park next to high-status cars or vice versa, and (2) how close do they park?

As the students spent time in parking lots and near businesses, watching cars being parked, they recorded data on a sheet similar to that in Table 8–10. The primary car is the car already parked and the secondary car is the one being parked next to the primary car. The status of the car, high or low, was not filled in at the same time the other data was collected. To provide uniformity in rating cars as high- or low-status, another group of people rated all cars after it was determined what cars had been included in the study. These twelve raters were given only the list of cars, and the results of their ratings were added to the data sheets. For data entry into the computer, each make, model, and color was given a code number, such as: Olds Cutlass Black might be coded "17" for Olds, "05" for Cutlass, and "01" for black. This would appear in the computer record as: 17 05 01. Year, status, and distance were added to the computer record as they appeared on the data sheet, with two numbers for year and two numbers for distance in inches. Year and distance are continuous data items; status, make, model, and color are labeled data items.

The cross-tabulation table will provide us with a grouping of the four possibilities for parking. The computer grouped all the P's for primary car together, then all the S's for secondary car, using the label that we had recorded for each car in the study. Within these groups all the H's for high status and all the L's for low status were grouped. The result is Table 8–11. It shows 19 high-status cars parked next to high-status cars, 17 high-status cars parked next to low-status cars, 43 low-status cars parked next to high-status cars, and 28 low-

Time ____________ Place ____________________

	Primary Car	Secondary Car
Status (H or L)	____	____
Make	____________	____________
Model	____________	____________
Color	____________	____________
Year	____________	____________
Distance between cars (inches)	______	

Table 8–10 Car Data Sheet

status cars parked next to low-status cars. The primary cars, cars already parked, included 62 high-status cars and 45 low-status cars. The secondary cars included 36 high-status cars and 71 low-status cars and the total numbers of pairs observed was 107. Low chi-square and high significance indicate no relationship, that is, parking habits are independent of car type with respect to a choice of parking space. The row, column, and total percentages normally associated with a cross-tabulation table were not reproduced in Table 8–11.

A scatter diagram (not reproduced here), along with correlation, was obtained to see if drivers of older cars had a tendency to park closer; that is, as the age of the car increased, did the distance between cars decrease? If true, this would produce a negative correlation. Correlation was .2, showing no relationship between age of car and distance left between cars. Age of car was determined by having the computer subtract each car year from the present year. As age and distance were the only two continuous data items, these were the only two for which we could obtain a scatter diagram.

For drivers of high-status cars choosing a parking space, the mean average distance between cars was 36½ inches; and for drivers of low-status cars choosing a parking space, the mean average distance between cars was 39 inches (low-status cars left more space, on the average).

Primary Car Status	Secondary Car Status: High	Low	Total
High	19	43	62
Low	17	28	45
Total	36	71	107

DF = 1
Chi-square = .32
Significance = .573

Table 8–11 Cross-tabulation—High-Status and Low-Status Cars

A frequency table was produced by make of car. As we had several groups of data gatherers, we did not expect any particular makes of cars to be observed disproportionately. Tables 8–12 indicates that our expectations were probably accurate. The two modes of 20 cars, with 9.3 percent each of the total cars, belong to Chevrolet (code 5) and Toyota (code 23), followed closely by Ford (code 9) with 19 cars observed, or 8.9 percent of the total of 214. The codes column

Makes	Codes	Frequency	Relative Frequency Percent
AMC	1	1	.5
BMW	2	8	3.7
Buick	3	14	6.5
Cadillac	4	13	6.1
Chevrolet	5	20	9.3
Chrysler	6	10	4.7
Datsun	7	14	6.5
Dodge	8	5	2.3
Ford	9	19	8.9
GMC truck	10	2	.9
Honda	11	5	2.3
Isuzu	12	1	.5
Lincoln	13	9	4.2
Mazda	14	1	.5
Mercedes	15	7	3.3
Mercury	16	4	1.9
Oldsmobile	17	16	7.5
Peugeot	18	1	.5
Plymouth	19	4	1.9
Pontiac	20	13	6.1
Porsche	21	5	2.3
Subaru	22	1	.5
Toyota	23	20	9.3
Triumph	24	1	.5
VW	25	14	6.5
Volvo	26	6	2.8
Total		214	100.0

Table 8–12 Frequency Table—All Cars by Make

shows the number assigned each make of car for ease of processing by computer. For your purposes, the makes of cars and matching codes would be recorded using the method provided by your particular statistics package (section 6.3). Recording the makes would be necessary only as an identifier for printing results; the computer would create groups using the code numbers. As there is no need for order here, the cumulative frequency column was not reproduced, and a mean, if requested, would be based on assigned code numbers, a useless number.

Primarily to satisfy curiosity, a frequency table was produced by age of car (mainly a factor in deciding status); a graph of this, grouped, is reproduced in Figure 8–10, with other statistics at the bottom of the figure that were calculated before the data was grouped.

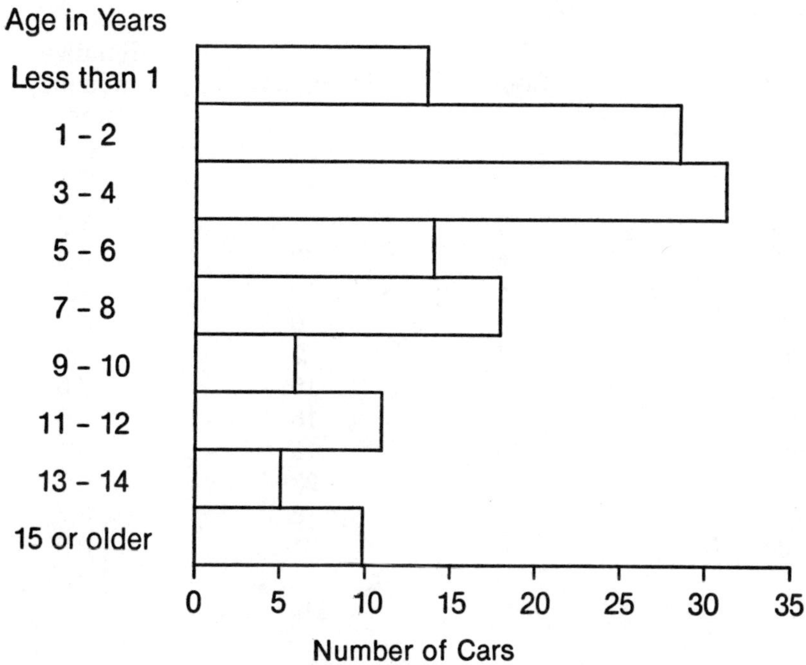

Mean	5.7	Minimum	0 or less than 1 year
Median	4.2	Maximum	22
Mode	2		

Figure 8–10 Bar Graph Age of Cars

The mean was 5.7 years old. The oldest car was 22 years. Two-year-old cars were observed most frequently, as denoted by the mode. A median of 4.2 means that about half the cars observed were 4 years or less and half were older than 4 years.

A frequency table by color of car (another factor in deciding status) was also produced. Table 8–13 shows blue to be the most popular color in this study, mode of 42 (21.8 percent). Brown (11.7 percent), white (10.7 percent), and green (9.3 percent) were observed quite a few times. Data gatherers failed to record color for 21 cars; this was coded as a "0" when data was entered for color of car, and it was counted as missing data. Body color was used in cases where body color and roof color were different.

The results of this study showed that we may be fooled by our observations; things are not always as they seem. In this case, parking habits are not related to the type of car being parked.

Color	Codes	Frequency	Relative Frequency Percent
Black	1	11	5.1
Blue	2	42	21.8
Brown	3	25	11.7
Green	4	20	9.3
Gray	5	10	4.7
Maroon	6	9	4.2
Orange	7	2	.9
Red	8	15	7.0
Silver	9	15	7.0
Tan	10	13	6.1
White	11	23	10.7
Yellow	12	8	3.7
Missing Data	0	21	9.8
Total		214	100.0

Table 8-13 Frequency Table All Cars by Color

9

Business and Industry

1. COMPARING THIS YEAR AND LAST YEAR

The information in Table 9–1 represents the number of units shipped to fifteen different customers or stores. To obtain the information the computer grouped all shipments by customer number, and month and year shipped, and totaled the units shipped for the appropriate period. We assume this is a part of your computer files. If units are not comparable, we can compare by costs or regroup units by unit ID number before doing the statistics. Studying the figures is not difficult for fifteen customers, but most businesses will have more than fifteen customers. Let's have the computer look first, and tell us if (1) there is a significant difference in our sales between last year and this year, or (2) we should investigate any particular customers. There is a significant difference between sets of comparable data if the means of the sets of data differ from each other more than might be expected by chance.

We have two pairs of numbers, continuous data, to compare—current month with previous year, current month, and then, year to date with previous year to date. For significant difference on a pair of numbers we use the t-test (section 5.2). The results of processing the data in Table 9–1 are shown in Table 9–2. The first column to consider is the last column, two-tail probability. If we want to be 95 percent certain that there is a significant difference between this year and last, then the last column figure should be .05 or less. For the year-to-date summary (YTD) this is true. The margin of error is less than 5 percent (.048). We can say, with 95 percent certainty, that there is a significant difference in number of shipments. For the monthly summary, we should not claim a significant difference, with a probability of .44. At least the mean average for this year is better in both cases, as seen by inspection of the mean or the mean differ-

Customer	Current Month	Previous Year Current Month	YTD	Previous YTD
1	874	0	1,448	0
2	0	15,932	0	15,932
3	21,263	17,779	37,974	36,064
4	699	0	1,874	0
5	3,450	1,140	4,818	1,959
6	2,096	2,002	3,589	3,714
7	1,088	626	3,104	1,205
8	1,015	732	3,130	1,817
9	4,935	0	4,935	0
10	5,513	0	10,035	0
11	13,230	6,281	22,372	15,053
12	11,954	12,591	25,557	13,147
13	11,892	11,547	32,973	13,369
14	14,493	7,778	38,139	16,657
15	1,892	1,547	6,369	2,973

Table 9-1 Units Shipped by Customer

	Number of Cases	DF	Mean	Standard Deviation	Mean Difference	r	t	2-tail Probability
Current Month	15	14	6293	6577	1096	.66	.79	.44
Prev. Yr. Cur. Mo.			5197	6354				
YTD	15	14	13088	14129	4962	.78	2.17	.048
Previous YTD			8126	10180				

Table 9-2 Pairs t-test—Number of Units Shipped

ence. If the mean difference were negative, then this year's average would have been worse than last year's. The same is true if the value for t were negative. The t value for YTD is considered high enough to claim a significant difference. R, or correlation, is positive, indicating that this is an appropriate use of the pairs t-test. Correlation of .78 is high enough to say, in most cases, that customers changed their orders uniformly—there were very few drastic increases or decreases. The standard deviation for this example is not much use for computing + or − 1 standard deviation. Subtracting the standard deviation from the mean, in this case, would result in an unreasonable figure of negative units shipped for all comparison periods. Degrees of freedom (DF) is n − 1, which is not important for our use, but is included because it is usually produced in computer results and mentioned in statistics books. The standard error may also be a part of the results on some computers.

Other useful statistics on this set of data are shown in Table 9-3 for the year-to-date figures. The 0 for minimum for previous year to date does not represent the same customer as that in the column, YTD as we omitted those customers not ordering for the past two years. Therefore, this previous year 0 indicates that we have gained at least one customer, one who did not order last year; we know we have lost at least one customer with the 0 for present YTD. This is shown with the minimum of 0 units shipped for present YTD. We know that more units were shipped this year to our best customer;

	YTD	Previous YTD
Minimum	0	0
Maximum	38139	36064
Median	4935	2973
Mean	13088	8126
S.D.	14129	10180
C.V.%	108	125
Skewness	0.9	1.6
Kurtosis	−0.8	2.8
Std.Error	3648	2628
95% C.I.	5263 − 20912	2489 − 13763

Table 9-3 Descriptive Statistics—Units Shipped

maximum this year was greater. There is more variation in last year's shipments, shown by the coefficient of variation, but there is a lot of variation both years. Skewness for both is positive and higher for the previous year—meaning that one, or a few customers, ordered more units than most other customers. However, the low value for skewness in both cases indicates that there were no large gaps in the number of orders between customers, particularly for this year. There is one outlier for the previous year. We saw this on customer 3 for previous YTD in Table 9-1. The means being larger than the median is in agreement with positive skewness. Kurtosis for previous YTD is positive and relatively high, meaning that there would be a peak in the data, if it were shown on a graph. Customer 3 would also have caused this. For this year, kurtosis is negative, and the graph would appear flatter; . . there are several customers with orders close to 30,000 units, so that the lengths of the bars would be nearly the same (this continuous data, however, is not normally shown by a bar graph). Using all of these comparative statistics we can claim that, in most cases, this year shows a trend toward customers' making our company their choice.

Table 9-1 was for data illustration; we don't want to pore over this type of list for all of our many customers. We want to see the best customers and those to whom we might want to give some extra attention (the worst customers). We might want to pick off those who ordered less than 1,500 units or more than 30,000 units, for example. We entered these two figures for YTD and obtained the listing in Table 9-4. If the list had been longer, we could have separated it by doing two picks—of those at less than 1,500 and then of those at more than 30,000. Using code matching (section 6.1) between the customer file and the inventory file, or a more comprehensive statis-

Customer	YTD
1	1448
2	0
3	37974
13	32973
14	38139

Table 9-4 Good and Poor Customers

tics package, we can receive the customer names, addresses, and any other necessary information.

These types of statistics can be used on any data for which you have a pair of comparable numbers, each of which has a label of some kind, such as employee number or department number. You might want to use the mean average and pick off those less than average. (Remember, if skewness is high positive, it will distort the mean, resulting in more customers' falling below the mean). For previous YTD there would have been nine customers, from Table 9–1, below the mean of 8,126. In this case, skewness is not high enough for the data to be extremely one-sided. In addition, we could pick off those who ordered less this year than last, and there are various other combinations. Also, appropriate business statistics can be used as proof of qualification for business loans, for example, trends showing increase in number of customers or number of sales.

A programmer doesn't have to write special programs to produce the many informative reports; a computer statistics package can do it all.

2. JOB DISSATISFACTION SURVEY

The objective in this study is to (1) determine some of the reasons people leave jobs, and (2) create a specific profile of turnover. We look at this survey from two angles—the characteristics of the person and the characteristics of the job. Surveying employees on the job will not always result in honest replies. A survey of employees after they have quit, and have nothing to lose, can give us some insight into reasons for a low retention rate, or it can give us some ideas we might not have considered previously about improving working conditions. Some managers have to be hit on the head to see that certain undesirable conditions exist, or to acknowledge the importance of trying to improve the conditions, if they are known. Other managers realize immediately that productivity, for example, could be improved if some of the day-to-day aggravations of workers could be removed, or if morale could be improved.

The following survey was conducted with hospital workers. Some of the questions may be different for other businesses, but the types of statistics used should be predominately the same. There is a more direct relationship to people-oriented jobs or to jobs with nonstandard working hours, such as shift work. A similar survey can also be used for education: why do students quit school or transfer colleges.

As these were prior employees, the questionnaires had to be mailed to those for whom addresses could be obtained. A stamped, self-addressed envelope was included to ensure a better return percentage. The survey form has not been reproduced in this section. The principles are the same as those in section 7.1. Questions are limited and allow a number to be circled. If instructions are not specific, the interpretations may differ widely. If a male answers yes to a female question, he has been given an opportunity to misinterpret the question; such things sometimes happen even when instructions *are* specific.

People are more likely to complete the questionnaire if it does not take much time. Space was allowed for comments but, normally people do not take the time to make comments. Comments cannot be used in computer statistics, will be different from the various respondents, and will probably not show a pattern. Even if someone has the time to read the comments, who can remember the first one by the time the fiftieth one is read? If another survey of the same type is planned and a good comment was found, then it can be included in question form the next time.

Names of respondents should not be requested. For certain types of surveys, an offer of a copy of the results will increase replies; a tear-off name and address space can be included. The respondent can then let you remove it after the survey is returned to you, or send it separately to ensure anonymity of the reply, if this is a factor in the survey. For this sensitive study of why people quit, a third party was used to gather the data.

When we are doing surveys by computer, some questions can be omitted; we can let the computer do them. For example, if we want to know age at resignation and how long it has been since the person worked in a hospital, we have to ask only one question. Assuming we've already asked for present age, the computer can subtract age at resignation from present age to determine how long it has been—or we can ask how long and subtract that from the present age to find age at resignation. (The time elapsed since a person worked in a hospital was of interest, as people might forget bad conditions after awhile; but that is a question beyond the scope of this book.)

After the data was entered into the computer, which took about five hours for 105 surveys, the first step was to get frequency tables on all the answers to see how the data looked. Five hours of data entry and a few minutes to produce statistics would be comparable to weeks or months of manual work. Each survey had ninety possible responses within twenty questions. Some questions required one response and some required several responses. Looking at the frequency tables will help us determine some of the other statistics we can use. For example, if fewer than five people replied yes to a yes-or-no question, then we don't want a cross-tabulation table for that question. As most of the questions were yes-or-no, we could see from the frequency tables if any questions were answered yes by all respondents or no by all respondents. No one claimed sexual discrimination as a reason for leaving a job; it was the only question answered no by all respondents.

Some of the usual questions were asked: age, marital status, sex, number of children, educational level, last job held in a hospital (staff or supervisory), length of time spent as a hospital employee, time elapsed since leaving, was the individual currently employed in a related job elsewhere, and did he or she ever plan a return to hospital work? Although determining the most important reasons for leaving was the primary purpose, many of the questions were slanted toward trying to determine people's attitudes, feelings, and perceptions of conditions during their last year of employment on that job.

Responses to these yes-or-no questions can be accumulated in frequency tables for the whole group and for subgroups, such as by educational degree, or last job, or time lapse grouping. We were interested in the percentage of people replying yes to questions. A report written about computer-produced statistics necessitates pulling out the important points and arranging them in whatever manner is appropriate for the information. This has been done in Figure 9–1 for the highest percentages of yes replies by the whole group.

The graph was made using the results of nine different frequency tables and looking at the results of other frequency tables in which the "yes" percentages and the "no" percentages were given for each question. The percentage claiming heavy workload indicates a problem of understaffing. This was the impetus for the survey in the first place—not having, or keeping, enough qualified employees. You can't please everyone on the second problem, hours or scheduling, but an

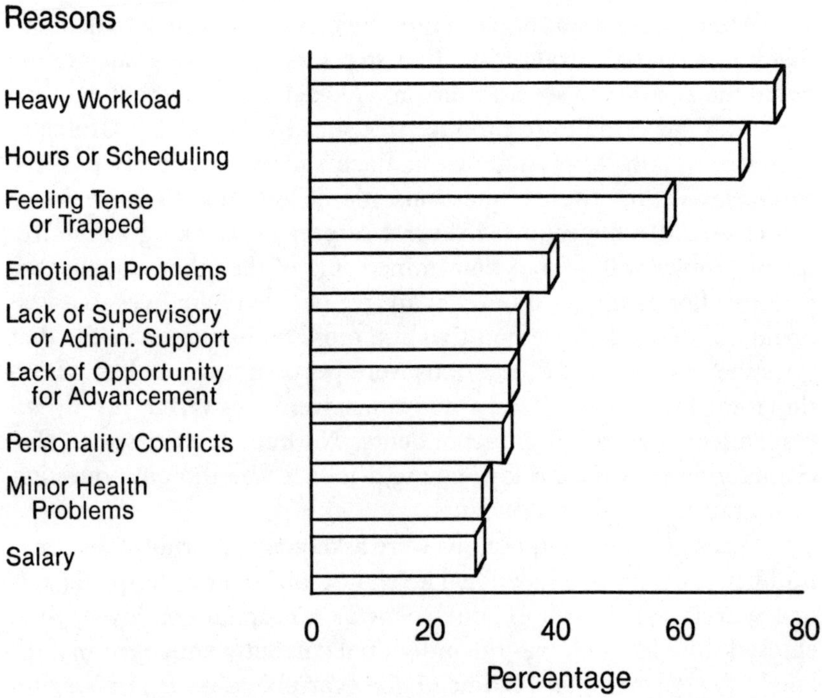

Figure 9-1 Bar Graph—Major Reasons for Leaving

attempt should be made to evaluate this for possible improvements. Some of the other reasons shown on the graph could be a direct result of the first two reasons for many employees. Lack of support or opportunity for advancement was checked by enough people to warrant an investigation. Salary was indicated by slightly more than a quarter of the respondents, maybe not an important issue for the group as a whole. Other reasons checked by less than a quarter of the respondents include: boredom, sleep and weight problems, family's health, major health problems, education not being utilized, feeling of isolation, benefits, increase in addictive habits, dependency on others, and ethical problems. The question was not asked, but some comments referred to too much "politics." It could not be counted in the statistical study because everyone did not have an opportunity to check this reason.

In glancing at the surveys as they were returned, we noticed that many people without degrees were in supervisory positions and many with degrees were in staff positions. We wondered if we were correct in this observation. A comparison of this labeled data can be achieved with a cross-tabulation table. According to Table 9–5, we might be right. The low chi-square and significance above .05, or even .10, tells us that we cannot claim a relationship between degree and position; position is not dependent on degree. Some doubt is present because of the small number of supervisors contained in the table. It might be argued that there are too many supervisors compared to staff (too many Chiefs and not enough Indians). However, that is not comparable data for our study; a larger percentage of supervisors may have responded to the survey than were working concurrently. Next, we split the data by age at resigning to see if age explained why those without degrees were in supervisory positions. The results still showed no relationship between degree and education. Age was not a factor. Having people with more education reporting to people with less education might be a problem that should be investigated.

	Highest Degree				
Position	Diploma	Associate Degree	Bacca-laureate	Master's Degree	Total
Staff	33 43 75 24	23 30 88 15	15 20 68 34	5 7 83 5	76 78
Supervisor	11 50 25 11	3 14 12 3	7 32 32 7	1 4 17 1	22 22
Total	44 45	26 27	22 22	6 6	98 100

DF = 3
Chi-square = 3.17
Significance = .37
Missing Data = 7

Table 9-5 Cross-tabulation Table—Degree and Position

In Table 9–5, the first number in each row is the actual number of respondents. The second number in each row is the row percentage and the third number is the column percentage. The last number is a percentage of the two rows and four columns; these last eight percentages will total 100 percent. Reviewing again how to interpret the numbers, using those with only a high school diploma, in the first column, we have: 33 were staff and 11 were supervisory, 43 percent of all the staff had only a diploma and 50 percent of all the supervisors had only a diploma, 75 percent of those with a diploma were staff and 25 percent of those with a diploma were supervisors. The outside totals figures, first going down and then across, show 76, or 78 percent, were staff; 22, or 22 percent, were supervisors; 44, or 45 percent, had a diploma; 26, or 27 percent, had an associate degree; 22, or 22 percent, had a baccalaureate degree; and 6, or 6 percent, had a master's degree. The closeness of the numbers with the percentages (rounded) is explained by the total number of respondents, 98. You know what would happen if we had 100 respondents answering these questions.

Comparisons of all the labeled, or count, data can be made using cross-tabulation tables: all of the yes-or-no responses in Figure 9–1 and those items listed but not included in Figure 9–1. Those answering yes and those answering no could be grouped by marital status, or sex, or highest degree, or position before resigning, or plans to return to hospital work, or number of children. Although number of children could be considered continuous data, the range was so small, 0 to 4, that it was considered labeled data; we could also obtain averages that made sense, whereas an average of the labeled data, marital status, does not make sense.

The data in Table 9–5 showed no relationship. This was true in most of the cross-tabulations when comparing the various factors. One condition found to have a relationship at the 97 percent level of confidence was marital status and claiming salary as a reason for leaving. This table is reproduced in Table 9–6, showing significance of .03. Looking at the second number in each cell: of those married, 22 percent left because of salary as one reason and 78 percent did not. Of those single, 54 percent left because of salary and 46 percent did not. An assumption is made that the salary might not be good enough for the single people. Looking at the third number in each cell: of those replying yes to salary as a reason for leaving, 74 percent

Marital Status	Salary as Reason for Leaving: Yes	No	Total
Married	20 22 74 19	72 78 92 68	92 88
Single	7 54 26 7	6 46 8 6	13 12
Total	27 26	78 74	105 100

DF = 1
Chi-square = 4.58
Significance = .03

Table 9-6 Cross-tabulation—Salary and Marital Status

were married and 26 percent were single. Of those replying no to salary as a reason for leaving, 92 percent were married and 8 percent were single. Before a final conclusion is made about this data, we should try to add more data on single people, or ensure before the survey that we have a representative sample. This is sometimes impossible with surveys of this type. If this is the case, we must avoid making claims about these unrepresentative items or state the inconclusive situations.

Other than number of children, continuous data items included were: present age, length of time spent working in a hospital, time elapsed since resignation, age at resignation, and the total number of reasons checked by each person. Some respondents checked only a couple of reasons, while others checked many reasons for leaving. With one command to the computer, these "yes" responses were counted and the total reasons checked were recorded with the other data for each person. All of this continuous data could be compared with the use of averages, or correlations, or the t-test, or analysis of variance. The t-test was appropriate for most of this since we had so much data that involved two groups—those replying yes and those replying no. Correlation could be used with any two factors, such as time elapsed since resignation and total number of reasons checked. The results of this showed very little correlation, −.34. The negative indicates that as the time lapse increased, the number of (remembered) reasons for leaving decreased.

T-tests were used on several factors. The only one in which we might claim a significant difference was one comparing position—staff or supervisory—and length of time employed in hospital work. This is shown in Table 9-7 as a groups t-test (Table 9-2 was a pairs t-test). The groups t-test is appropriate for comparing the same measurement on two groups of people exposed to different conditions; in this case, the conditions are staff and supervisory, and the measurement is length of time employed in hospital work. The average for staff was 5.4 years and the average for supervisors was 7.6 years. Generally, we keep supervisors longer than staff. The standard deviation tells us that the variation around the mean was about the same for both groups. About two-thirds (68 percent) of the staff stayed from 1.1 years to 9.7 years (5.4 + or −4.3) and of the supervisors from 3.6 years to 11.6 years (7.6 + or −4.0). We obtained no data on length of time worked before they became supervisors. The t value is negative because the second mean is larger than the first. According to the last figure for probability, we can be 94 percent sure that there is a significant difference in length of time worked between staff and supervisors. DF is n −2 for two groups.

	Number of Cases	DF	Mean	Standard Deviation	F	t	2-tail Probability
Staff	60	76	5.4	4.3	1.2	−1.86	.06
Supervisor	18		7.6	4.0			

Missing Data = 27

Table 9-7 Groups t-test—Years in Hospital Work by Position

Reasons given for resigning could not be associated with any particular group. In most cases, the whole group seemed to dislike the same conditions, regardless of sex, age, or any other category. Many times a survey, or parts of a survey, can be as valuable in showing what *is not* true as it is in showing what *is* true. A much more in-depth study was done than has been given here, or than is needed for this book. The final aim was to develop a profile of turnover. As stated before, statistics give us a point at which to start looking.

It is suggested that with a study of this magnitude, you decide before you request the statistics what statistics you want, matched up with the type of data. It is much easier to review the results if they are grouped thus: all frequency tables together, all cross-tabulation tables together, all correlations together, etc. It is confusing to look at one frequency table, then a cross-tabulation table, then another frequency table, then something else. The computer will produce the statistics in any order you wish; make it easy on yourself after the fact by thinking before the fact.

To help you think ahead, we will repeat a few general suggestions to use as a guide in running surveys that will be processed by computer (some are true if a computer is not used).

1. Keep questions simple and limited to the necessary information.
2. Request answers in numeric form, multiple-choice whenever possible.
3. Provide a number to circle or to fill in on a mark-sense form (section 6.2).
4. Number each and every question for easy identification or reference to a specific question. Even if a question has more than one part, each part should have a unique sequential number.
5. Be consistent with expected rating responses from high to low, or from low to high; do not use both in the same survey. Too many choices for one question will confuse the respondent.
6. Do not ask unnecessary questions when the computer could provide the answer by combining other questions or using simple arithmetic (also a time-saver for manual data entry).
7. Do not use leading questions or implications that inadvertently prompt a certain answer.
8. Be specific with instructions.

9. Be careful about the order in which questions are asked; previous questions can influence a person's chain of thought, causing related subsequent questions to receive biased answers.
10. For data entry, use an identification number for each respondent in the event that specific data has to be corrected or added to.
11. If manual data entry is necessary, provide a form or a simple method for the data-entry person; for example, by each question on one of the blank answer sheets, designate the column of the computer record where each particular answer is to be located. The columns in the computer record or the definitions of the data items should be in the same order as that of the questions on the survey, if this method is used. Include the possible numeric codes for each response in a question if they are not a part of the question.
12. Provide a consistent number or numbers to denote missing data (unanswered questions).
13. For a better return percentage on mail-in surveys, include a stamped self-addressed envelope and a promise of a copy of the results, if wanted, or information about publication of the results. It is imperative that this promise be kept.
14. If necessary, before developing the survey, get some help from people knowledgeable about arranging questions and data in a form suitable for the computer. (The help might be free).

3. THE SALESPERSON'S CUSTOMER RECORDS

If you are in a position to find characteristics and facts about customers helpful in sales, you can use your experience, or imagination, about what these facts might include for you. In the world of competition, having more knowledge about the product, combined with knowledge about people, can put you a step ahead of the competition. Very few businesses can make use of the "puppy dog" method (telling customers to take it home for a while and see how they like it); and high-pressure methods may produce sales if a satisfied customer is not important.

The need for more specific knowledge is seen in the personal record of one real estate salesperson. It is an attempt to answer some questions with recorded facts rather than relying on memory.

1. Under what conditions do I need to work harder?
2. When should I use encouragement without appearing to use pressure?
3. How can I take advantage of past history about what people actually buy and what they say they want?
4. What types of real estate are giving me the most sales?

In addition to the usual questions of name, address, phone, occupation, and possibly recreational interests, the items in Table 9-8 are entered into the computer about each contact. For simplicity, it is not all-inclusive—and we will not include the many nonstatistical possibilities with automation by computer to produce greater sales for the salesperson such as, mailing lists of prospective customers and selectively printing follow-up letters or matching people's desires with the features of houses for sale.

Property owner referral under *Source of Contact* means referral by those home owners who have bought from this real estate company. The *Drive-in* is a reference to chance contacts—people who "were out driving around and looking." Each of the major headings in Table 9-8 will have a space in the computer record and the appropriate codes will be recorded; for example, a 3 in the space for location desired will denote interior lot. Although these records are meant to be a help to a competent salesperson, they could also be combined with other sales records for use by the company. Then we could have a comparison of salespersons and the characteristics of their success, or failure.

Another code on each record (not shown in Table 9-8) denoted three different categories: prospects, sales, or no interest (bought elsewhere, or no longer worth our time). The last category can be used to see if there is a pattern to our failures. Computer statistics packages will not provide for permanent updating of the customer file. This would have to be done with other software; for example, the code for prospects will eventually become the code of sales or no interest. Assuming the information has been updated, we can use the file to compare all of the information in Table 9-8 for the whole group, or for the different categories, which can be combined or picked off by code number.

All the data is labeled with the exception of prices. Number of children is continuous data with a limited number of values. The dates can produce continuous data by computing times—how long it takes

Date of Initial Visit ______________

Source of Contact or Interest
1. Media
2. Property Owner Referral
3. Non-property Owner Referral
4. Other Real Estate Broker
5. Drive-in

Location
1. Waterfront
2. Golf Course
3. Interior

Lot Type
1. Large Lot
2. Small Lot

Structure Type
1. House
2. Condo
3. Patio Villa

Reason for Interest or Purchase
1. First Home
2. Second Home
3. Retirement Home
4. Investment

Age
1. Under 35
2. 35 to 49
3. 50 to 65
4. Over 65

Number of Children __________

Property Purchase Information
- ID Number __________
- List Price __________
- Selling Price __________
- Contract Date __________
- Closing Date __________

Table 9-8 Customer Contact or Sales Records

to make a sale (e.g., difference between date of initial visit and closing date). This has been covered in section 7.1 with reference to hours or days spent in jail. Other continuous data can be produced with the difference in list price and selling price. The mode of the month portion of the dates can tell us our busiest month of the year or our most lucrative month (commissions might not come in the month the sale was made).

First we will concentrate on prospects versus sales. Figure 9-2 is a comparison of what people claimed they wanted and what they actually bought with reference to location. In all three cases it is quite different. Figure 9-3 is the same with reference to structure type. The people buying patio villas appear to be more certain about what they want, or what they can afford.

The location compared to reason for purchase was of special interest to this particular real estate salesperson. In most cases throughout this section, we are interested in percentages only. We will con-

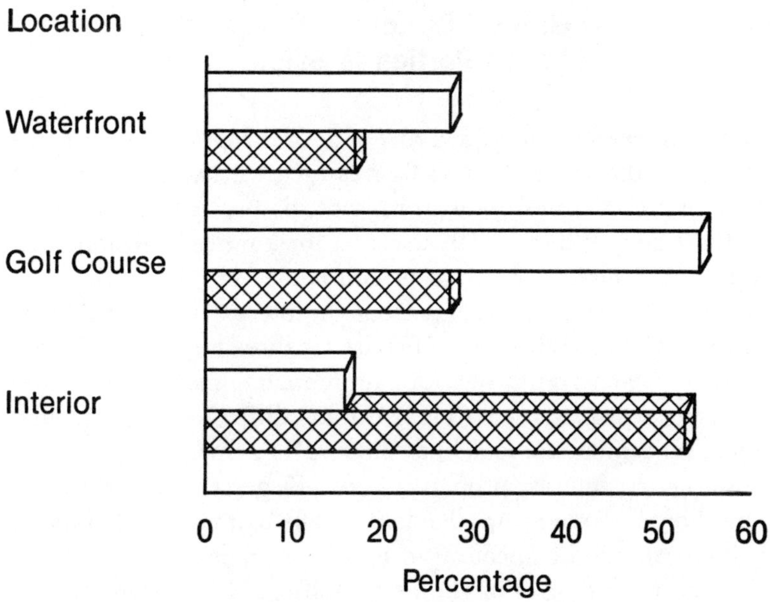

Figure 9-2 Bar Graph—Location: Prospects Versus Sales—Shaded Portion Is Sales

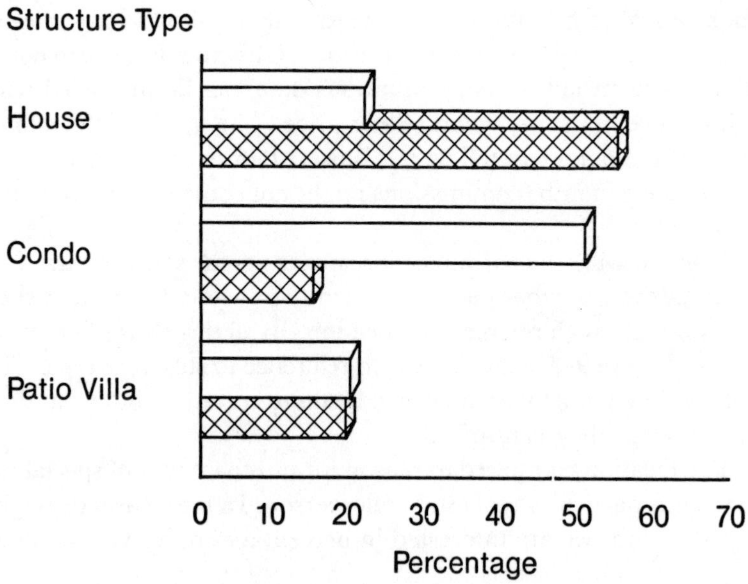

Figure 9-3 Bar Graph—Structure: Prospects vs. Sales Shaded Portion is Sales

centrate on the second and third row of figures reproduced in Table 9-9. Reading the second row of figures for each, we see: for those buying a first home, most bought interior, 63 percent; for a second home, most bought interior, 91 percent; for a retirement home, an equal percent bought waterfront or golf course property, 37.5 percent; for investment purposes, most bought interior property, 71 percent. Looking at the third row of figures for those buying waterfront property, all, 100 percent, were for a retirement home; for those buying golf course property, there is not much difference between first home and retirement home, 34 percent and 33 percent, respectively; for those buying interior property, most, 45 percent, bought it for a second home. To sum up, Table 9-9 tells us that people buying a first home should be encouraged to look at interior property also even though their desires might be for golf course; those wanting a retirement home should be encouraged to look at both waterfront and golf courses regardless of what they say they want, and so forth. The same comparisons can be made with the other labeled data in Table 9-8 at the beginning of this section. Using past history on sales might

Reason for Purchase	Location: Waterfront	Location: Golf Course	Location: Interior	Total
First Home	0 0 0 0	3 37 34 9	5 63 23 15	8 24
Second Home	0 0 0 0	1 9 11 3	10 91 45 29	11 32
Retirement Home	3 37.5 100 9	3 37.5 33 9	2 25 9 6	8 24
Investment	0 0 0 0	2 29 22 6	5 71 23 14	7 20
Total	3 9	9 27	22 64	34 100

Table 9-9 Cross-tabulation—Buyers by Location and Reason for Purchase (Data changed for illustration)

lessen the possibility of wasting our time and the customers' time by showing them everything.

All of the tables will not be reproduced, but some of the strengths and weaknesses of this salesperson can be extracted from various types of comparisons. Two of these are shown in Table 9-10 and 9-11. In Table 9-10, the second row for property owners, 18 percent of the total buyers were from those referred by a property owner and 9 percent of the total contacts were referred by a property owner. This is a compliment to the salesperson and the real estate company—that satisfied customers are producing a proportionally larger number

Source of Contact	Buyers	Total Contacts
Media	9%	13%
Property Owner	18%	9%
Nonproperty Owner	20%	21%
Other Broker	29%	24%
Drive-in	24%	33%
	100%	100%

Table 9-10 Source of Contact—Buyers vs. Total Contacts

Age	Buyers	Total Contacts
Under 35	18%	13%
35 to 49	35%	27%
50 to 65	35%	45%
Over 65	12%	15%
	100%	100%

Table 9-11 Age—Buyers vs. Total Contacts

of sales (a good source of free advertising). This is true also for referrals by other brokers, 29 percent and 24 percent. The sales produced by the media or drive-ins resulted in a lower percentage of sales compared to the percentage of contacts. This has implications for our advertising and for staffing the office waiting for drive-ins. Time data, by day, or week, or month, might be useful. We could continue this investigation by having the computer separate the data by the different sources of contact and make comparisons of those items in Table 9-8. In Table 9-11, it can be seen that this salesperson is more successful with the under-50 age group (the first two rows), a larger percentage of buyers compared to contacts.

Finally, we will create a short profile of the buyers by extracting the largest percentages from the various tables produced by the computer. This can be done with all pertinent areas. We will use age, a factor in many statistical studies. The under-35 group was referred by a property owner, and bought a house on a large interior lot as a first home. The 35-to-49 age group was referred by a nonproperty owner, and bought a house on a large interior lot as a second home. The 50-to-65 age group was referred by another broker, and bought a patio villa on a small golf course lot as a retirement home. The over-65 age group was informed by the media, and bought a house on a large interior lot as an investment. Information of this kind can also be beneficial in planning subdivisions and directing advertising.

In these examples and other examples throughout the book, some of the results may not be what we would expect, nor what we would guess if we depended upon our memory, or our own perception of situations.

Statistics start with a need to find out something. Think ahead about what information you need, thus avoiding the time-consuming job of entering all kinds of information and receiving useless reports; then be sure that all the necessary data is saved in the computer. If the statistics are wrong, then the data was wrong. Using a computer can be compared to having your picture taken—don't complain when you don't get any more out of it than you put into it. What you put into or get out of the computer should be consistent every time. In order to discover patterns or trends, the same items of data must be compared. For a simple example, if we are comparing month and number of contacts over the year, we should not switch our comparisons to month and number of sales for one of the months. This is a different data item, a different comparison, and a different pattern indicator. (We would then have the proverbial "apples and oranges" comparison.) We should not jump from one track to another, to another, until we have decided definitely that we are going in the wrong direction.

You have been given some ideas about what kinds of information would be useful for you, and what kinds of things you can do with the information you have, especially if you reread the book with this in mind. As you have seen, statistics programs have other uses besides producing statistics (in the usual mathematical sense). The picking and sorting (rearranging) and grouping features can provide us with many facts about data, even if our knowledge of statistics is minimal.

Getting accurate information might mean the difference between success and failure. In years to come, computer statistics will become easier and easier to use; large data bases of information will become more accessible to more people. In years to come, we may all have the same access to the information we need; then, the difference will be, not in the machinery, or in the information it produces, but in the people—the salesperson, the manager, the teacher—and their ability to *use* the information.

Glossary

Analysis of variance: a test of differences between two or more sets of continuous data grouped by a label, such as educational level, or income categorized.

Anova: see *Analysis of variance.*

Ascending order: arrangement of numbers in a list in order from low to high, or of letters from A to Z.

Bar graph: a frequency table shown in graph form as a set of vertical or horizontal bars. The lengths of the bars indicate the relative size of each group of values. Used with labeled data or categorized continuous data.

Chi-square: a number showing if data is independent or dependent (related). Large chi-square implies that a relationship of some kind exists; small chi-square implies that data is independent. Used with labeled data or categorized continuous data.

Code: an alternate name for a data item, or for the values a data item might include, such as marital status—0 = single, 1 = married, 2 = widowed.

Coefficient of determination: see *R squared.*

Coefficient of variation: the amount of variability in a set of data as a percentage; the standard deviation divided by the mean. The larger the number, the more variability. Used with continuous data.

Computer statistics package: see *Statistics package.*

Confidence Interval (95%): the range within which the mean of related data will be expected to fall 95 percent of the time; + or −2 standard errors. Used with continuous data. See also *Standard error.*

Constant: a number or label that does not change, that has a fixed value.

Contingency table: see *Cross-tabulation table.*

Continuous data: measurement data such as age, physical distance; continuous data can have arithmetic used on it that can provide a decimal result.

Correlation: the measurement of the relationship between two qualities or sets of data. Used with continuous data. Also see r, positive correlation, and negative correlation.

Cross-tabulation table: the combining of two or more individual frequency tables into a frequency array to compare the frequency distribution of the items in one table with the items in the other tables; shows a trend. Used with labeled data or categorized continuous data.

Cumulative frequency: the addition, one by one, of each individual percentage in the relative frequency column of a frequency table. The last percentage listed should be 100 percent.

Data: a record of individual or collective information; also a set of related information.

Degrees of freedom: related to the number of cases and/or number of groups; more than one or two numbers are needed to compute statistics, resulting in degrees of freedom of $n - 1$, or $n - 2$, or similar expressions; implies freedom to vary.

Dependent variable: a value or measurement which is determined or predicted depending on the value of another variable.

Descending order: arrangement of numbers in a list in order from high to low, or of letters from Z to A.

Descriptive statistics: those that describe a set of data, usually including minimum, maximum, range, mean, mode, median, standard deviation, variance, skewness, kurtosis. Used with continuous data (or labeled data for mode).

Explained variation: see *R squared.*

F statistic: a number showing the difference between two or more sets of continuous data. The larger the F value and the smaller the significance, the more confident we are that there is a difference between the sets of data.

File: a set of related information or records stored together in a computer.

Frequency distribution: see *Frequency table.*

Frequency table: a list of values with the count of occurrences of each value or the count within each group of values. The values are arranged in ascending order or descending order. Used with labeled data or with continuous data with a limited number of values.

Histogram: a graph of a frequency table using vertical bars. The length of the bars indicates the relative size of each group of values (the number of occurrences of a value within each group). Used with labeled data or categorized continuous data.

Independent variable: a value or measurement that can be used to predict another value; usually comes first in time.

Interaction: the effect one or more factors have on other factors in analysis of variance, such as educational level on attitude ratings.

Intercept: the value of y when x is 0 in the regression equation. Also called y-intercept or "a." Used with continuous data. See also *Regression equation*.

Kurtosis negative: no concentration of data around certain values; when shown in a graph, the graph will appear flat. Used with continuous data.

Kurtosis positive: data will be concentrated around certain values; when shown in a graph, the graph will have a peak or peaks. Used with continuous data.

Labeled data: data that is categorized, or the value is a label or name, such as sex, race, birthplace.

Level of significance: see *Significance*.

Linear relationship: pairs of numbers on a scatter diagram form a straight line (or an approximate straight line for statistical use). Used with continuous data.

Mark-sense form: special paper for gathering data by filling in boxes or circles.

Mark-sense reader: a machine that reads pencil marks on special forms and enters the data into the computer.

Maximum: the highest value in a set of data. Used with numbered data.

Mean: an average obtained by summing a set of values and dividing by the total number of values. Used with continuous data.

Mean difference: the number obtained when subtracting the mean of one group or set of data from the mean of another group or set. Used with continuous data.

Measurement data: see *Continuous data*.

Median: an average obtained by arranging a set of data in order and finding the middle value, or a number halfway between two middle values. Used with continuous data.

Menu: the list of choices that allows us to select, by number, the programs to be run by the computer; usually displayed on a screen by number and name of option. See also *Software*.

Minimum: the lowest value in a set of data. Used with numbered data.

Mode: the value that appears most frequently in a set of data; the value or the values with the largest count. Can be used with continuous data, but is more often used with labeled data.

Negative correlation: relationship between pairs of values; the approximate straight line will slope downward from left to right in a scatter diagram; as x increases, y decreases. Used with continuous data. See also *R*.

Normal distribution: most of the data is concentrated around the mean with less data as values get farther from the mean; a symmetrical graph with a peak at the mean and sloping down toward both ends. Used with continuous data.

One-tail probability: a term meaning the margin of error can be in one direction only, expressed as a decimal or percent. Used with continuous data.

Outliers: one, or a few extreme data values, on either end, after a large gap in the data values. Used with continuous data. See also *Skewness*.

Percentile: indicates where one measurement stands in relation to the group of measurements—percentage of measurements that fall below a particular measurement. Used with continuous data.

Positive correlation: relationship between pairs of values; the approximate straight line will slope upward from left to right in a scatter diagram; as x increases, y increases. Used with continuous data. See also *R*.

Probability: see *Significance, One-tail probability*, and *Two-tail probability*.

R: correlation coefficient between −1 and +1 showing the relationship between two sets of data. The closer to −1 or to +1, the more relationship. Used with continuous data.

R squared: a number between 0 and 1 showing how much of a relationship in correlation is due to the factors we are comparing; .7 means 70 percent of variation in dependent variable is caused by independent variable. Used with continuous data.

Range: the difference between the lowest value and the highest value in a set of data. Used with numbered data.

Record: all pertinent information about one person or item stored together in a computer.

Relative frequency: the percentage of occurrences of each value or group of values in relationship to other values in a frequency table. Used with continuous data or labeled data. See also *Frequency table, Percentile.*

Regression equation: $y = a + bx$, the prediction equation for linear correlation; y is the dependent variable we are predicting, x is the independent variable we are basing our prediction on, b is slope or the change in y for every unit change in x, a is the constant or intercept (the value of y when x is 0). Used with continuous data.

Regression line: an imaginary line that could be drawn through a set of points plotted on a scatter diagram. Used with continuous data.

Scatter diagram: a graph showing pairs of numbers, each of which meets at a point opposite one of the pair on a vertical axis and the other on a horizontal axis. Used with continuous data.

Significance: odds of a result happening by chance; .05 means the results would happen by chance only 5 out of 100 times, .01 only once in 100 times, .001 only once in 1,000 times, etc.

Significant difference: sets of comparable data differ from each other more than might be expected by chance. See also *Significance.*

Skewness negative: a few extremely low values in a set of data. The larger the value for skewness, disregarding negative, the more extreme the values are in relation to the rest of the data. Used with continuous data.

Skewness positive: a few extremely high values in a set of data. The larger the value for skewness, the more extreme the values are in relation to the rest of the data. Used with continuous data.

Slope: the change in y for every unit change in x in the regression equation. Also called "b." Used with continuous data. See also *Regression equation*.

Software: a program or set of programs that provide operating and data processing instructions to the computer.

Standard deviation: a measure of variability in a set of comparable data; the larger the standard deviation, the more variability. For normal large samples, about 68 percent, or 2/3's, of the cases will be within +1 or −1 S.D.; about 95 percent will be within 2 standard deviations; and practically all within 3 standard deviations. Can be obtained by taking the square root of the variance. Used with continuous data.

Standard error (of mean): the "fudge" factor in predicting a mean; the range within which the mean of related data will be expected to fall. Used with continuous data.

Standard error of estimate: the "fudge" factor in predicting with regression; the range within which the prediction should fall. Used with continuous data.

Statistics package: a computer program or set of programs to produce statistics.

Sum of squares: the number obtained after subtraction, squaring, and adding a set of numbers two by two; normally, but not exclusively, associated with variance. Used with continuous data.

System (as used in this book): a series of computer programs used for updating individual records and obtaining reports from sets of related data; for example, an accounting system or a students' information system.

T statistic: a number showing the difference between two sets of continuous data. The larger the t value and the smaller the significance, the more confident we are that there is a difference between the two sets of data.

T-test: a test of difference between two sets of continuous data, grouped by a label such as sex, or paired by a label such as before-and-after.

Two-tail probability: a term meaning the margin of error is plus or minus, expressed as a decimal or percent. Used with continuous data. See also *Significance.*

Unexplained variation: see *Standard error of estimate.*

Updating: adding, changing, or deleting information in a computer file record.

Variability: amount of deviation, or distance from a mean, or a standard (expectation). Used with continuous data.

Variable: a number or measurement or label that varies or changes with different circumstances, that has no fixed value.

Variance: a measure of variability around the mean of a set of data; obtained by subtracting each value in the set from the mean, one by one, squaring, and adding the results. Used with continuous data. See also *Standard deviation* and *Sum of squares.*

Variation: see *Variability.*

X-axis: the horizontal line in a scatter diagram; contains the independent variable values. Used with continuous data.

Y-axis: the vertical line in a scatter diagram; contains the dependent variable values. Used with continuous data.

Y-intercept: see *Intercept.*

Z-score: an expression of standard deviation with the mean assigned a value of zero. Used with continuous data.

INDEX

D

H

I

K

L

M

N

O

P

T